Handbook of the Mammals of Europe

Series Editors

Klaus Hackländer
Department of Integrative Biology and Biodiversity Research
Institute of Wildlife Biology and Game Management
University of Natural Resources and Life Sciences, Vienna (BOKU)
Vienna, Austria

Frank E. Zachos
Mammal Collection
Natural History Museum Vienna
Vienna, Austria
Department of Evolutionary Biology
University of Vienna
Vienna, Austria
Department of Genetics
University of the Free State
Bloemfontein, South Africa

This handbook offers a unique collection of information on all wild mammal species living in Europe and will serve as a standard reference guide for all mammalogists and readers interested in research on mammals. The introductory volume provides an overview of mammalian phylogeny, the history and current status of European mammals and their management as well as habitats, and the history of mammalogy in Europe. The remaining volumes present comprehensive species-specific chapters covering all aspects of mammalian biology, including paleontology, physiology, genetics, reproduction and development, ecology, habitat, diet, mortality, and behavior. The economic significance and management of mammals and future challenges for research and conservation are addressed as well. Each chapter includes a distribution map, a photograph of the animal, and a list of key literature. This authoritative handbook provides current and detailed descriptions of all European mammals; it will appeal to academics and students in mammal research alike as well as to practitioners whose work involves mammal management, control, use, and conservation.

You can find more information about this series at https://www.springer.com/series15198.

Klaus Hackländer • Frank E. Zachos
Editors

Mammals of Europe - Past, Present, and Future

With 16 Figures and 7 Tables

Springer

Editors
Klaus Hackländer
Department of Integrative Biology
and Biodiversity Research
Institute of Wildlife Biology
and Game Management
University of Natural Resources
and Life Sciences, Vienna (BOKU)
Vienna, Austria

Frank E. Zachos
Mammal Collection
Natural History Museum Vienna
Vienna, Austria

Department of Evolutionary Biology
University of Vienna
Vienna, Austria

Department of Genetics
University of the Free State
Bloemfontein, South Africa

ISBN 978-3-030-00280-0 ISBN 978-3-030-00281-7 (eBook)
ISBN 978-3-030-00282-4 (print and electronic bundle)
https://doi.org/10.1007/978-3-030-00281-7

Cover illustration: Cover photos with kind permission from:
Frank Zachos (European hamster, Red fox)
Ingo Arndt (Young European hares)

This Springer imprint is published by the registered company Springer Nature Switzerland AG.
The registered company address is: Gewerbestrasse 11, 6330 Cham, Switzerland

Acknowledgments

We are grateful to the editorial team of Springer Nature, especially to Barbara Wolf, Veronika Mang, Verena Penning, and Lars Körner. In addition, we thank our volume editors and all species chapter authors for their support and contributions.

Contents

About the Editors

Klaus Hackländer studied Zoology at Philipps-University Marburg/Lahn and the University of Vienna. He is currently a Full Professor of Wildlife Biology and Wildlife Management at the University of Natural Resources and Life Sciences, Vienna, where he also coordinates the master's degree program on Wildlife Ecology and Wildlife Management. Hackländer is a member of numerous mammalogical societies and served as president of the German Society for Mammalian Biology from 2009 to 2016. He was Editor-in-Chief of the journal *Mammal Review* from 2009 to 2015 and has coedited two books on the Lagomorpha, his favorite mammalian taxon. Hackländer's research covers all aspects of mammalian biology, including ecology, behavior, physiology, diseases, and management.

Frank E. Zachos is Head of the Mammal Collection at the Natural History Museum in Vienna, Austria, an Affiliated Lecturer at the University of Vienna, and an Affiliated Professor in the Department of Genetics, University of the Free State in Bloemfontein, South Africa. He studied Biology, the History of Science, and Philosophy in Kiel and Jena, Germany, and received his Ph.D. in Zoology in 2005. In 2009, he completed his habilitation (postdoctoral degree) in Zoology and Evolutionary Biology. His main research interests are intraspecific biodiversity, phylogeography, population genetics, and conservation of mammals and birds, with a particular focus on red deer and other ungulates. He also has a long-standing interest in species concepts and the species problem as well as other theoretical and philosophical issues in evolutionary biology and systematics. He is the Editor-in-Chief of the journal *Mammalian Biology* and the Editor of the Mammalia series within the *Handbook of Zoology*.

Contributors

Karl-Georg Bernhardt Department of Integrative Biology and Biodiversity Research, Institute of Botany, University of Life Sciences and Natural Resources, Vienna (BOKU), Vienna, Austria

Klaus Hackländer Department of Integrative Biology and Biodiversity Research, Institute of Wildlife Biology and Game Management, University of Natural Resources and Life Sciences, Vienna (BOKU), Vienna, Austria

Rainer Hutterer Leibniz Institute for Animal Biodiversity, Zoologisches Forschungsmuseum Alexander Koenig, Bonn, Germany

Boris Kryštufek Slovenian Museum of Natural History, Ljubljana, Slovenia

Robert S. Sommer Department of Landscape Sciences and Geomatics, Neubrandenburg University of Applied Sciences, Rostock, Germany

Department of Zoology, Institute of Biosciences, University of Rostock, Rostock, Germany

Arie Trouwborst Tilburg Law School, Department of Public Law and Governance, Tilburg University, Tilburg, The Netherlands

Frank E. Zachos Mammal Collection, Natural History Museum Vienna, Vienna, Austria

Department of Evolutionary Biology, University of Vienna, Vienna, Austria

Department of Genetics, University of the Free State, Bloemfontein, South Africa

The New *Handbook of the Mammals of Europe*: Background and Introduction

1

Klaus Hackländer and Frank E. Zachos

Contents

These are good times for mammalogy. New mammal species are constantly being described, and a renewed interest in taxonomy and biodiversity in general has resulted in a number of authoritative new book series on mammals, most notably the *Handbook of the Mammals of the World* (since 2009, e.g., Wilson and Mittermeier 2009) but also a new (first?) volume of the famous *Walker's Mammals of the World* (Nowak 2018). There are also continent-specific multivolume publications like the *Mammals of Africa* (Kingdon et al. 2013) and the *Mammals of South America* (so far two volumes: Gardner 2007; Patton et al. 2015), and it therefore seems timely to also tackle the task of an updated *Handbook of the Mammals of Europe*. The most comprehensive such series to date is the multivolume *Handbuch der Säugetiere Europas*. Its first volume was published in 1978 (Niethammer and Krapp 1978; Fig. 1) and the last, an overall index and bibliography, in 2005. None of the earlier volumes has ever been updated, and the fact that the whole series is in German makes it largely inaccessible to the vast majority within the European mammalogical community and beyond. This is exactly where the present book series comes in. We aim to present an updated account in English of every living mammal species in Europe. As some original research is still published in non-English journals, we provide an extensive and (near-)complete compilation of current knowledge about each mammal species in Europe.

The present introductory volume aims at setting the stage for the core volumes dealing with

K. Hackländer
Department of Integrative Biology and Biodiversity Research, Institute of Wildlife Biology and Game Management, University of Natural Resources and Life Sciences, Vienna (BOKU), Vienna, Austria
e-mail: klaus.hacklaender@boku.ac.at

F. E. Zachos (✉)
Mammal Collection, Natural History Museum Vienna, Vienna, Austria

Department of Evolutionary Biology, University of Vienna, Vienna, Austria

Department of Genetics, University of the Free State, Bloemfontein, South Africa
e-mail: frank.zachos@nhm-wien.ac.at

K. Hackländer, F. E. Zachos (eds.), *Mammals of Europe - Past, Present, and Future*, Handbook of the Mammals of Europe, https://doi.org/10.1007/978-3-030-00281-7_1

Fig. 1 Cover and title page of the first volume of the *Handbuch der Säugetiere Europas*, published in 1978. (Courtesy of Quelle & Meyer Verlag, 7 Oct 2019)

mammalian diversity in Europe at the species level. The selection of chapters in introductory volumes is often, perhaps by necessity, somewhat arbitrary, and we could have included other topics in addition to the ones dealt with here, for instance, reviews on diseases and zoonoses or cultural aspects of mammals throughout European history, i.e., their spiritual role in mythology and different societies, and their depiction in the arts from the awe-inspiring cave paintings by our ancestors to Rilke's poem *The Panther* and beyond. What we have decided on, apart from this extended preface, is five chapters that hopefully provide a framework and perspective for the species accounts in the remaining volumes.

Because geology, climate, and vegetation differ hugely across Europe, the contribution by Karl-Georg Bernhardt deals with the continent's different biogeographical regions, with a particular focus on vegetation ▶ Chap. 4, "Mammal Habitats in Europe: Geology, Vegetation, and Climate". This chapter serves as a reference and background for the information in the species accounts on habitat and distribution, especially for readers outside Europe.

Frank Zachos gives a brief summary and overview of the substantial progress made over the last 20 or so years with respect to mammalian phylogenetics ▶ Chap. 3, "Mammalian Phylogenetics: A Short Overview of Recent Advances". Although the whole taxon Mammalia is covered, including monotremes and marsupials, the focus is on placental mammals, in accordance with the extant European mammal fauna. Today, mammals in Europe belong to a limited number of higher taxa ("orders") – rodents, lagomorphs, a single (nonhuman) primate species, eulipotyphlans (i.e., what is left of the former "Insectivora"), bats, carnivorans, and cetartiodactyls including whales and dolphins. In the past, however, a number of groups today perceived as exotic also occurred in Europe, among them proboscideans (most famously, the woolly mammoth), great apes, pangolins, and even marsupials. A wider phylogenetic perspective therefore seemed appropriate.

The less distant past of mammalian faunas is covered by Robert Sommer's chapter on the Late Quaternary history of mammals in Europe (▶ Chap. 5, "Late Pleistocene and Holocene History of Mammals in Europe"). The cyclic climate changes brought about by glacial and interglacial periods have had a deep impact on the biogeography of mammals and other taxa, and any present distribution and composition will always be the result of past processes.

In addition to the history of mammals themselves, there is another chapter, by Rainer Hutterer and Boris Kryštufek, on the history of research on mammals in Europe. Mammals, even European ones, include many iconic and emblematic species such as European bison (*Bison bonasus*), red deer (*Cervus elaphus*), wolves (*Canis lupus*), or whales, and apart from birds, there are few other animal groups that have fascinated humans as much as mammals have. Accordingly, there is a long tradition of mammal research in Europe and a large number of mammalogical societies and publications (including journals), a brief overview of which is presented in ▶ Chap. 2, "A History of Mammal Research in Europe" by Hutterer and Kryštufek.

Finally, since globally about one in four (perhaps even one in three) mammal species is threatened with extinction and more than half of the world's mammals are declining (Schipper et al. 2008; see also Turvey 2018), conservation and sustainable use of mammals is key to their survival. On the other hand, conflict species including aliens are controlled to reduce ecological or economic damage. Klaus Hackländer and Arie Trouwborst provide an introduction to policies, laws, and strategies pertaining to the management of mammals in Europe, focusing in particular on the pan-European Convention on the Conservation of European Wildlife and Natural Habitats (the Bern Convention of 1979) and the wildlife legislation of the European Union ▶ Chap. 6, "Management of European Mammals".

The *Handbook* at large mainly consists of species entries, combined into volumes covering larger taxonomic groups such that, including the present introduction volume, the content looks like this:

1. Mammals of Europe – Past, Present, and Future
2. Primates and Lagomorpha
3. Rodentia
4. Eulipotyphla
5. Chiroptera
6. Carnivora
7. Terrestrial Cetartiodactyla
8. Cetacea

Some of these will perhaps comprise more than one tome. The list of volumes follows phylogenetic relationships within mammals but also makes compromises: (1) the single European primate species, the Barbary macaque (*Macaca sylvanus*), is combined with the lagomorphs on account of their both being part of the larger taxon Euarchontoglires. (2) Although it is now well-established that Cetacea (whales and dolphins) are nested within artiodactyl ungulates – hence the name Cetartiodactyla for the taxon combining both – and even that they are phylogenetically closer to the ruminant clade than are the pigs, we still divide the Cetartiodactyla into a terrestrial and an aquatic group. We are aware that the terrestrial cetartiodactyls are paraphyletic, but there has been a long research tradition to study cetaceans separately from their terrestrial relatives, and many issues pertaining to the European ruminant species (Cervidae and Bovidae) will also be relevant to the wild boar but not whales and dolphins.

In line with other handbooks and reference works, we have decided on a standardized chapter structure with (largely) fixed subheadings to enable direct comparisons between different species. Of course, some species have been studied in much more detail than others, but differences in content and depth among species when it comes to, say, the Life history or the Genetics sections ideally reflect the real differences in our knowledge. Still, in some cases, there are deviations from the strict standardized framework. In the cetacean volume, for example, a slightly modified chapter structure has been adopted in line with the very different research circumstances pertaining to whales and dolphins. For instance, the Genetics section has been combined with information on abundance into a section called Populations. Also, there is a single chapter on rare, vagrant, and extinct cetaceans rather than a chapter for each such species.

In the case of introduced mammal species that have very limited distribution ranges in Europe (e.g., axis deer (*Axis axis*) or Reeves's muntjac (*Muntiacus reevesi*)), the chapters are not only short but may also deviate slightly from the structure found for other species entries. The case of feral domestic taxa is also quite complex at times. In general, we have decided against including chapters on feral cats or horses. In some cases, however, it is not so easy to decide whether we are dealing with ancient feral domestic or truly wild taxa. We followed most authors in treating wild sheep and goats as at least potentially wild European mammal species.

Geographic Delimitation

Europe as a continent is obviously a political construct. From a geographic point of view, it is a rather arbitrary entity and quite simply the westernmost part of Eurasia. Therefore, the spatial delimitation of the region covered by a handbook such as ours is not straightforward, and there are different ways in which Europe can be defined spatially. Following the approach adopted by the National Geographic Society, we consider as Europe's western and southern boundaries the Ural Mountains, the Caucasus Mountains, and the Bosphorus, respectively. We also include Cyprus, the Canary Islands, Madeira, and the Azores as well as other islands that belong to a European country's European territory (e.g., Svalbard in the case of Norway). In contrast to the distribution maps of *The Atlas of European Mammals* (Mitchell-Jones et al. 1999), we therefore include countries like Belarus, Ukraine, parts of Russia (including Novaya Zemlya and Franz Joseph Land), and Moldova (the revised Atlas, to be published in 2024, will also include the eastern European countries). However, we do not take overseas departments and regions of France (French Guiana, Guadeloupe, Martinique, Mayotte, and Réunion) or British Overseas Territories (Bermuda, Cayman Islands, Falkland Islands, etc.) into consideration, with the exception of Gibraltar due to its location on the Iberian Peninsula. Different maps of Europe in mammal publications and identification guides might be confusing, but as mentioned above, Europe is a political construct and not a biogeographic continent. The boundaries of Europe as defined for the purpose of the present *Handbook* can be seen in the map in Fig. 2.

A related issue is which species should be included in a handbook of *European* mammals. When it comes to whales, which are able to cross whole ocean basins, there is inevitably some level of arbitrariness (and there will be a chapter on rare and vagrant species, see above), but the ranges of a number of Asian species also extend into the margins of what we have defined as Europe. We have decided not to include primarily Asian mammals (e.g., the saiga antelope, *Saiga tatarica*). Nor did we include chapters on extinct species such as lions (*Panthera leo*) (but extinct cetaceans will be briefly covered together with the rare and vagrant species). In the case of roe deer, we only cover the European species (*Capreolus capreolus*), while the Siberian roe deer (*C. pygargus*) is only briefly mentioned in the *C. capreolus* chapter where relevant.

Taxonomic Approach

A handbook of all species of a region de facto includes a taxonomic list of that region. As every biologist knows, there are different such lists

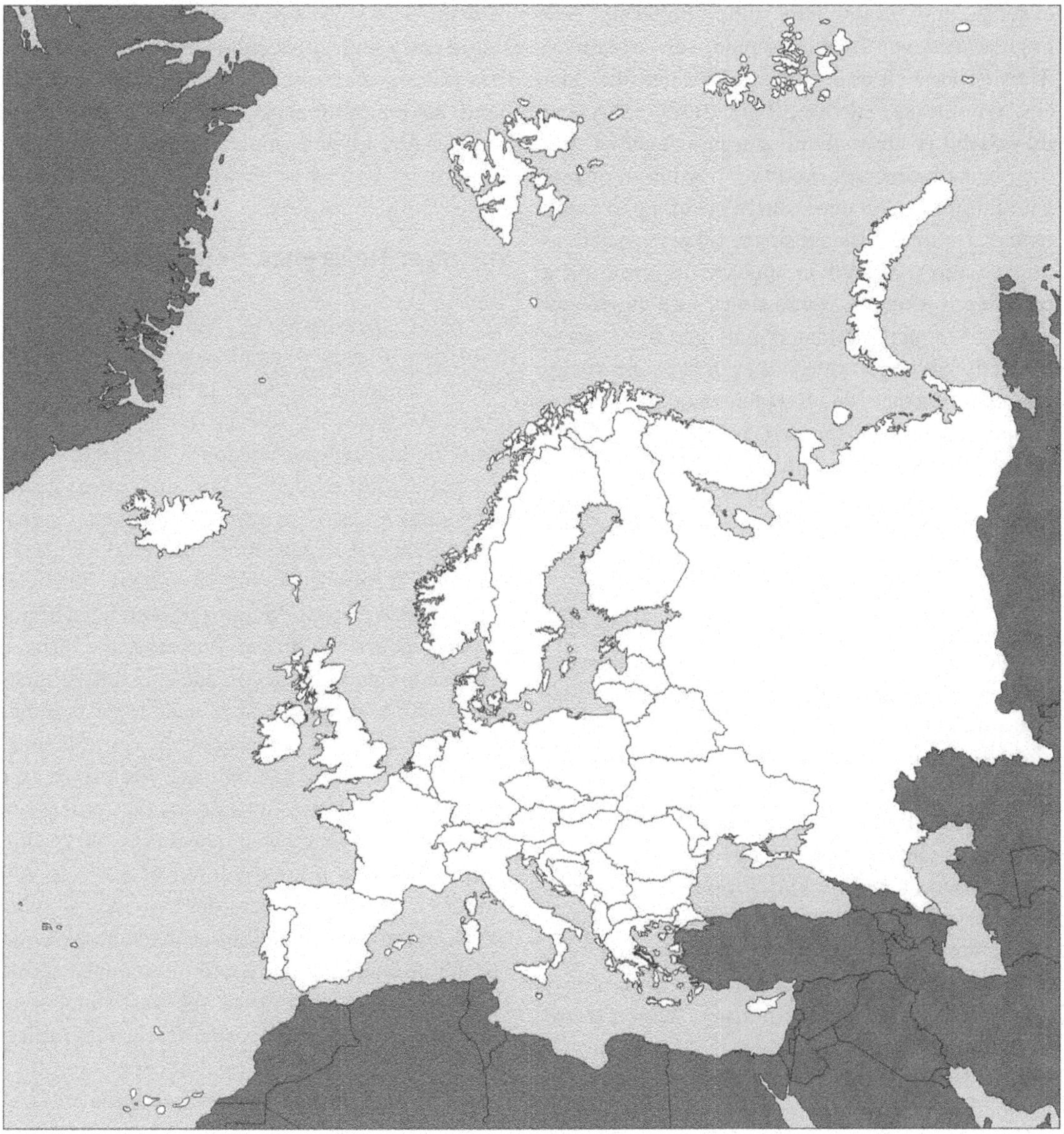

© Getty Images/iStockphoto

Fig. 2 Europe (in white) as defined in the present *Handbook of the Mammals of Europe*

available for a given region that are accepted by different taxonomists from different taxonomic schools of classification. Therefore, any decision on species delimitation will necessarily be appreciated by some and criticized by others. Mammals have occupied center stage in recent debates on species concepts and species delimitation, and there have been many recent publications that deal with the underlying theory and its application in taxonomic practice (e.g., Frankham et al. 2012; Groves 2013; Heller et al. 2013; Zachos et al. 2013; Zachos 2018; Gippoliti 2019, and references therein). We will not discuss these issues at length or repeat the major arguments in favor of more or less inclusive species concepts. Rather, we would like to emphasize that species delimitation necessarily contains an element of arbitrariness – where exactly one draws the line between two population-level lineages needs an executive decision as well as it does good scientific data and hypotheses (Zachos 2016). Biodiversity and its variability are real and can be measured, but they come in degrees, and the translation of real data into names is necessarily arbitrary to some extent. Taxonomy is a discrete binary classification system (one or two species) imposed on a continuous process (evolution), and there can never be a perfect match and certainly not a perfectly objective one either. There are many mammal species in Europe that have been lumped or split, re-lumped, or re-split. We and the volume editors had to make a decision as to which taxa are granted species rank in our *Handbook* and which are not. Are the *musculus* and *domesticus* taxa of the house mouse (*Mus musculus*) subspecies or species? If we followed a strict phylogenetic species concept based on diagnosability (which we don't), even red deer in Europe would comprise several different species, not subspecies (in addition to *Cervus elaphus* there would, at least, also be *C. pannonensis*, *C. corsicanus*, and *C. italicus*). Other formerly more inclusive species are now usually classified as different species, among them different *Myotis* and *Pipistrellus* species. Other cases in which hitherto unknown ("cryptic") diversity has recently been uncovered have not been settled by the taxonomic community (e.g., the hazel dormouse *Muscardinus avellanarius*; see Mouton et al. 2017). In any case, a handbook is not the place for detailed taxonomic discussions or even revisions, and volume editors and authors have some level of liberty in their taxonomic treatments. This is why complete consistency among volumes will probably not be reached, which should be considered a reminder of inherent problems in taxonomy due to fuzzy boundaries in nature. However, taxonomic traditions also have practical ramifications in that for some taxa that are now considered different species a large part of the literature does not necessarily distinguish between them, making it difficult to treat them in separate chapters. This is, for example, the reason why the two chamois species (*Rupicapra rupicapra* and *R. pyrenaica*) are combined into a single entry with subsections. Alpine and Iberian ibex (*Capra ibex* and *C. pyrenaica*), on the other hand, have a more separate research history and are therefore given completely separate chapters.

Springer Reference Benefits

A major drawback of the *Handbuch der Säugetiere Europas* has already been mentioned above: language barriers. In present times an important factor in science is accessibility. In line with this, open-access journals emerge everywhere allowing academics and students to gather knowledge as easily as never before (even if keeping a clear view gets more difficult). Through Springer Reference, our *Handbook* provides online access as well. Mammal researchers, libraries or universities might purchase the printed series or single volumes. In addition, all volumes are available in digital form, and single volumes or species chapters are accessible via university libraries or other institutions. Moreover, the online system allows for a permanent process of updating and revising chapters as necessary, without publishing a revised edition of a whole volume. Our *Handbook* therefore provides a living source of knowledge on European mammals without major language barriers but with easy accessibility. We are grateful to the vast number of colleagues sharing their expertise in this project and thank Springer Nature Publishers for their trust, patience, and professional editorial work.

Acknowledgments The authors wish to thank Hermann Ansorge, Görlitz, for helpful comments on an earlier version of this chapter.

References

Frankham R, Ballou JD, Dudash MR, Eldridge MDB, Fenster CB, Lacy RC et al (2012) Implications of different species concepts for conserving biodiversity. Biol Conserv 153:25–31

Gardner AL (ed) (2007) Mammals of South America. Volume 1: marsupials, xenarthrans, shrews, and bats. The University of Chicago Press, Chicago/London

Gippoliti S (2019) Species delimitation in mammals: a comment on Zachos (2018). Mamm Biol 94:127–131

Groves CP (2013) The nature of species: a rejoinder to Zachos et al. Mamm Biol 78:7–9

Heller R, Frandsen P, Lorenzen ED, Siegismund HR (2013) Are there really twice as many bovid species as we thought? Syst Biol 62:490–493

Kingdon J, Happold D, Butynski T, Hoffmann M, Happold M, Kalina J (eds) (2013) Mammals of Africa (6 volumes). Bloomsbury Publishing, London

Mitchell-Jones AJ, Amori G, Bogdanowicz W, Kryštufek B, Reijnders PJH, Spitzenberger F, Stubbe M, Thissen JBM, Vohralík V, Zima J (1999) The atlas of European mammals. T & AD Poyser, London

Mouton A, Mortelliti A, Grill A, Sara M, Kryštufek B, Juškaitis R et al (2017) Evolutionary history and species delimitations: a case study of the hazel dormouse, *Muscardinus avellanarius*. Conserv Genet 18:181–196

Niethammer J, Krapp F (eds) (1978) Handbuch der Säugetiere Europas. Band 1, Nagetiere I. Akademische Verlagsgesellschaft, Wiesbaden

Nowak RM (2018) Walker's mammals of the world. Monotremes, marsupials, afrotherians, xenarthrans, and sundatherians. Johns Hopkins University Press, Baltimore

Patton JL, Pardinas FJ, D'Elía G (eds) (2015) Mammals of South America. Volume 2: Rodents. The University of Chicago Press, Chicago/London

Schipper J, Chanson JS, Chiozza F, . . . Young BE (2008) The status of the world's land and marine mammals: diversity, threat, and knowledge. Science 322:225–230

Turvey ST (2018) Mammal extinction risk and conservation: patterns, threats, and management. In: Zachos FE, Asher RJ (eds) Mammalian evolution, diversity and systematics. Handbook of zoology, Mammalia. de Gruyter, Berlin/Boston

Wilson DE, Mittermeier RA (eds) (2009) Handbook of the mammals of the world. Volume 1: Carnivores. Lynx Edicions, Barcelona

Zachos FE (2016) Species concepts in biology. Historical development, theoretical foundations and practical relevance. Springer International Publishing, Basel

Zachos FE (2018) Mammals and meaningful taxonomic units: the debate about species concepts and conservation. Mammal Rev 48:153–159

Zachos FE, Apollonio M, Bärmann EV, Festa-Bianchet M, Göhlich U, Habel JC, Haring E, Kruckenhauser L, Lovari S, McDevitt AD, Pertoldi C, Rössner GE, Sánchez-Villagra MR, Scandura M, Suchentrunk F (2013) Species inflation and taxonomic artefacts – a critical comment on recent trends in mammalian classification. Mamm Biol 78:1–6

A History of Mammal Research in Europe

2

Rainer Hutterer and Boris Kryštufek

Contents

Abstract

A review is given on the historical development of mammal research in Europe. The term Mammalia was coined by Carolus Linnaeus in 1758 for animals bearing mammae. While he named 77 species, a current list counts 320+ species for Europe. Thirty-five journals specialized on mammals have been issued since 1926, about 22 of which are currently being published. Regional mammal congresses have been organized since 1926, European-wide ones since 1960. Mammal Societies have been founded in Germany (1926), The Netherlands (1952), the United Kingdom (1954), France (1954), Czechoslovakia (1958), Italy (1983), Lithuania (1989), Russia (1992), Ukraine (1993), and Spain (2000), among others. Local faunas and handbooks have been published in many different countries over the last 250 years, culminating in the present *Handbook of the Mammals of Europe* and in the *Handbook of the Mammals of the World*.

R. Hutterer (✉)
Leibniz Institute for Animal Biodiversity, Zoologisches Forschungsmuseum Alexander Koenig, Bonn, Germany
e-mail: r.hutterer@leibniz-zfmk.de

B. Kryštufek
Slovenian Museum of Natural History, Ljubljana, Slovenia
e-mail: bkrystufek@pms-lj.si

K. Hackländer, F. E. Zachos (eds.), *Mammals of Europe - Past, Present, and Future*, Handbook of the Mammals of Europe, https://doi.org/10.1007/978-3-030-00281-7_5

Keywords

Mammal research · History · Theriology · Taxonomy · Nomenclature · Europe

Introduction

Mammalogy is a branch of zoology dealing with mammals. The term blends Mammalia (i.e., mammals) and -logy (from Ancient Greek logos: a principle of knowledge, a reason, or a study). The hairy animals with a chain of three bones in the middle ear, a single bone in the lower jaw which articulates with the squamosal bone, the left aortic arch of the fourth pharyngeal arch, and with cheek-teeth bearing several roots, were known to early naturalists as Theria (from therion, Greek for a wild beast) or Quadrupedia (a combination from quadri, to mean four, and pes for hand, both Latin; i.e., quadrupeds or four-handed). The term Mammalia was coined by Linnaeus in the 10th edition of Systema Naturae (1758), to denote "these and no other animals [which] have mammae [mammata]" (Schiebinger 1993). Mamma (plural is mammae) is a Latin term designating the milk-secreting organs of females and translates either as breast or teat. The term mammalogy (also mammology) therefore literally means "a study of breasts/teats" and not of breast-bearing animals (Schiebinger 1993). The term mastology, which is in use in the Portuguese-speaking World (e.g., Brazil), has identical connotations, meaning a study of mammary glands. Another term was coined by Pallas (1811), a Russian naturalist of German origin, namely, Lactantia (i.e., alluding to breast-feeding).

Mammalogy explores every aspect of structure, function, and natural history of mammals and incorporates diverse aspects of management of wild populations. Usually, mammalogy focuses on free-living mammals, both extant and fossil, but leaves domesticated forms to veterinary medicine and animal husbandry. Scientists who study mammals may be primarily interested in them per se or may utilize these animals as models to understand more general biological principles. Depending on this, students of mammals can identify themselves as mammalogists, or as ecologists, population biologist, behavioral scientists, physiologists, conservationists, morphologists, wildlife managers, paleontologists, evolutionary biologists, and so forth. In consequence, one can trace scientific papers dealing with mammals in a broad spectrum of periodicals, starting from journals which specialize on mammals, to periodicals covering nearly any field of biology.

The diversity and complexity pose problems also to students of the history of mammalogy. We therefore restricted ourselves to the activities focusing on delimitation of mammalian species and documenting mammal faunas. In its narrow sense, these are the fields of taxonomy and zoological nomenclature. These two fields together with descriptive morphology and rudimentary zoogeography formed the beginnings of modern mammalogy.

Europe as a Starting Point

The formal establishment of zoological nomenclature by the Swede Carolus Linnaeus (1707–1778) can be regarded as the starting point of systematic mammal research in Europe (Linnaeus 1758). His work, however, was based on numerous earlier publications by uncounted authors, such as Gessner and Forer (1563), which are not further treated here. Linnaeus named 77 European mammal species (see Table 1), most of which are still valid. The study of mammal taxonomy continues until today. Reasons are the changing techniques such as the study of chromosomes and DNA sequences which allow deeper insight into the speciation process, in different views on delimitation of species, but also the study of the last remaining unexplored spaces in Europe.

Most early researchers dealing with mammals came from Europe. The term mammalogy was introduced (as French mammologie) by a French zoologist Anselme-Gaëtan Desmarest (1784–1838) in 1820. In comparison with ornithology, which was in usage already in the sixteenth century, the term mammalogy emerged relatively late and was also hesitantly applied during the nineteenth century when ornithology was already widely used. In the early twentieth century, mammalogy was still only occasionally used in French

Table 1 Preliminary list of recognized mammal species known to occur in Europe as defined by Hackländer and Zachos (this volume) with authors, year, and source. Note that this list is not identical with the eventual list of species chapters in this handbook.

Primates
Macaca sylvanus (Linnaeus, 1758). Syst. Nat., 10th ed. 1:25
Homo sapiens Linnaeus, 1758. Syst. Nat., 10th ed. 1:20
Lagomorpha
Lepus capensis Linnaeus, 1758. Syst. Nat., 10th ed. 1:58
Lepus castroviejoi Palacios, 1977. Donana, Acta Vertebrata 1976, 3(2):205
Lepus corsicanus de Winton, 1898. Ann. Mag. Nat. Hist. ser. 7, 1:155
Lepus europaeus Pallas, 1778. Nova. Spec. Quad. Glir. Ord. p. 30
Lepus granatensis Rosenhauer, 1856. Die Thiere Andalusiens 3
Lepus timidus Linnaeus, 1758. Syst. Nat., 10th ed. 1:57
Oryctolagus cuniculus (Linnaeus, 1758). Syst. Nat., 10th ed. 1:58
Sylvilagus floridanus (J. A. Allen, 1890). Bull. Amer. Mus. Nat. Hist. 3:159
Soricomorpha/Eulipotyphla
Atelerix algirus (Lereboullet, 1842). Mem. Soc. Hist. Nat. Strasbourg, 3(2), art. QQ:4
Erinaceus concolor Martin, 1838. Proc. Zool. Soc. London 1837:103(1838)
Erinaceus europaeus Linnaeus, 1758. Syst. Nat., 10th ed. 1:52
Erinaceus roumanicus Barrett-Hamilton, 1900. Ann. Mag. Nat. Hist. ser. 7, 5:365
Hemiechinus auritus (Gmelin, 1770). Nova Comm. Acad. Sci. Petropoli 14:519
Crocidura canariensis Hutterer, López-Jurado & Vogel, 1987. J. Nat. Hist. 21:1354
Crocidura gueldenstaedtii (Pallas, 1811). Zoogr. Rosso-Asiat. 1:132
Crocidura leucodon (Hermann, 1780). In Zimmermann, Geogr. Gesch. Mensch. Vierf. Thiere 2:382
Crocidura pachyura (Küster, H.C., 1835). Isis von Oken 28:77[75–78]
Crocidura russula (Hermann, 1780). In Zimmermann, Geogr. Gesch. Mensch. Vierf. Thiere 2:382
Crocidura sicula Miller, 1900. Proc. Biol. Soc. Wash. 14:41
Crocidura suaveolens (Pallas, 1811). Zoogr. Rosso-Asiat. 1:133
Crocidura zimmermanni Wettstein, 1953. Z. Säugetierk. 17:21
Diplomesodon pulchellum Lichtenstein, 1823. Eversmann, Reise von Orenburg nach Buchara, p. 124
Suncus etruscus (Savi, 1822). Nuovo Giorn. De Letterati, Pisa 1:60
Neomys anomalus Cabrera, 1907. Ann. Mag. Nat. Hist. ser. 7,20:214
Neomys milleri Mottaz, 1907. Mém. Soc. Zool. France 20:22
Neomys fodiens (Pennant, 1771). Synopsis Quadrupeds p. 308
Neomys teres Miller, 1908. Ann. Mag. nat. Hist. 1:68
Sorex alpinus Schinz, 1837. Neue Denkschr. Allgem. Schweiz. Gesell. Naturwiss. Neuchatel 1:13
Sorex antinorii Bonaparte, 1840. Iconogr. Faun. Ital. 1:29
Sorex araneus Linnaeus, 1758. Syst. Nat., 10th ed. 1:53
Sorex averini Zubko, 1937. Kharkov A. Gorsky-State Univ., Proc. Zool.-Bot. Inst. 4:300
Sorex caecutiens Laxmann, 1788. Nova Acta Acad. Sci. Petropoli 1785, 3:285 (1788)
Sorex coronatus Millet, 1828. Fauna de Main-et-Loire I, p. 18
Sorex granarius Miller, 1910. Ann. Mag. Nat. Hist. ser. 8,6:458
Sorex isodon Turov, 1924. C. R. Acad. Asci. Paris, p. 111
Sorex minutus Linnaeus, 1766. Syst. Nat. 12th ed. 1:73
Sorex minutissimus Zimmermann, 1780. Geogr. Gesch. 2:385
Sorex raddei Satunin, 1895. Arch. Naturgesch. 1:109
Sorex samniticus Altobello, 1926. Bol. Inst. Zool. Univ. Roma 3:102
Sorex satunini Ognev, 1922. Ann. Mus. Zool. Acad. St. Pétersb. 22:331
Sorex tundrensis Merriam, 1900. Proc. Wash. Acad. Sci., 2: 16
Sorex volnuchini Ognev, 1922. Ann. Mus. Zool. Acad. St. Pétersb. 22:322

(continued)

Table 1 (continued)

Desmana moschata (Linnaeus, 1758). Syst. Nat., 10th ed. 1:59
Galemys pyrenaicus (E. Geoffroy St. Hilaire, 1811). Ann. Mus. Hist. Nat. Paris 17:193
Talpa aquitania Nicolas, Martínez-Vargas & Hugot, 2017. Mammalia 81:641
Talpa caeca Savi, 1822. Nuovo Giorn. de Letterati Pisa 1:265
Talpa caucasica Satunin, 1908. Mitt. Kaukasus Museum 4:5
Talpa europaea Linnaeus, 1758. Syst. Nat., 10th ed. 1:52
Talpa levantis Thomas, 1906. Ann. Mag. Nat. Hist. ser. 17:416
Talpa martinorum Kryštufek, Nedyalkov, Astrin & Hutterer, 2018. Bonn. zool. Bulletin 67:45
Talpa occidentalis Cabrera, 1907. Ann. Mag. Nat. Hist. ser. 7, 20:212
Talpa romana Thomas, 1902. Ann. Mag. Nat. Hist. ser. 7, 10:516
Talpa stankovici V. Martino & E. Martino, 1931. J. Mammal. 12:53
Chiroptera
Rousettus aegyptiacus (E. Geoffroy, 1810). Ann. Mus. Natn. Hist. Nat. Paris 15:96
Rhinolophus blasii Peters, 1867. Monatsber. K. Preuss. Akad. Wiss. Berlin 1866:17
Rhinolophus euryale Blasius, 1853. Arch. Naturgesch. 19(1):49
Rhinolophus ferrumequinum (Schreber, 1774). Die Säugethiere 1:174, pl.62
Rhinolophus hipposideros (Bechstein, 1800). In Pennant, Allgemeine Uebersicht Vierfüss. Thiere 2:629
Rhinolophus mehelyi Matschie, 1901. S.B. Ges. Naturf. Berlin, p. 225
Tadarida teniotis (Rafinesque, 1814). Précis. Som., p. 12
Eptesicus bottae Peters, 1869. Monatsber. K. Preuss. Akad. Wiss. Berlin 1869:406
Eptesicus anatolicus Felten, 1971. Senckenbergiana biol. 52:371
Eptesicus isabellinus (Temminck, 1840). Monogr. Mammalogie 2:205, pl. 52, f.1,2
Eptesicus nilssoni (Keyserling & Blasius, 1839). Arch. Naturgesch. 5(1):315
Eptesicus serotinus (Schreber, 1774). Die Säugethiere 1:167
Nyctalus azoreum (Thomas, 1901). Ann. Mag. Nat. Hist. ser. 7, 8:34
Nyctalus lasiopterus (Schreber, 1780). In Zimmermann, Geogr. Gesch. Mensch. Vierf. Thiere 2:412
Nyctalus leisleri (Kuhl, 1817). Die Deutschen Fledermäuse, Hanau p. 14, 46
Nyctalus noctula (Schreber, 1774). Die Säugethiere 1:166
Pipistrellus hanaki Hulva & Benda, 2004. Acta Chiropterlogica 6:193–217
Pipistrellus kuhlii (Kuhl, 1817). Die Deutschen Fledermäuse, Hanau p. 14
Pipistrellus maderensis (Dobson, 1878). Cat. Chiroptera Brit. Museum: 231
Pipistrellus nathusii (Keyserling & Blasius, 1839). Arch. Naturgesch. 5(1):320
Pipistrellus pipistrellus (Schreber, 1774). Die Säugethiere 1:167
Pipistrellus pygmaeus (Leach, 1825). Zool. J. 1:559
Barbastella barbastellus (Schreber, 1774). Die Säugethiere 1:168
Plecotus auritus (Linnaeus, 1758). Syst. Nat., 10th ed. 1:32
Plecotus austriacus (J. Fischer, 1829). Synopsis Mamm. p. 117
Plecotus begognae de Paz, 1994. Mammalia 58:423–432
Plecotus kolombatovici Đulić, 1980. Proc. 5th Internat. Bat. Res. Conf. (Wilson D E & Gardner A L eds), Texas Tech Press p. 159
Plecotus sardus Mucedda, Kiefer, Pidinchedda & Veith, 2002. Acta Chiropterol. 4:123
Plecotus teneriffae Barrett-Hamilton, 1907. Ann. Mag. Nat. Hist. ser. 7, 20: 520
Plecotus macrobullaris Kuzjakin, 1965. In Bobrinskij, Kuznetsov & Kuzyakin (Eds) Opredelitel mljeko-pitayushtshikh SSSR (p. 99). Moskva: Izd. Prosveshtshenije
Hypsugo savii (Bonaparte, 1837). Fauna Ital. 1, fasc. 20
Vespertilio murinus Linnaeus, 1758. Syst. Nat., 10th ed. 1:31
Myotis alcathoe Helversen & Heller, 2001. Naturwissenschaften 88:217
Myotis bechsteinii (Kuhl, 1817). Die Deutschen Fledermäuse, Hanau p. 14, 30
Myotis brandtii (Eversmann, 1845). Bull. Soc. Nat. Moscow 18(1):505

(continued)

Table 1 (continued)

Myotis capaccinii (Bonaparte, 1837). Fauna Ital., 1 fasc. 20
Myotis crypticus Ruedi, Ibánez, Salicini, Juste & Puechmaille, 2019. Acta Chiropterol. 20:291
Myotis dasycneme (Boie, 1825). Isis Jena, p. 1200
Myotis daubentonii (Kuhl, 1817). Die Deutschen Fledermäuse, Hanau p. 14
Myotis davidii (Peters, 1869). Mber. Preuss. Akad. Wiss. 1869:402
Myotis emarginatus (E. Geoffroy, 1806). Ann. Mus. Natn. Hist. Nat. Paris 8:198
Myotis escalerai Cabrera, 1904. Mem. Soc. Esp. Hist. Nat. 2(5):249–286
Myotis myotis (Borkhausen, 1797). Deutsche Fauna 1:80
Myotis mystacinus (Kuhl, 1817). Die Deutschen Fledermäuse, Hanau p. 15
Myotis nattereri (Kuhl, 1817). Die Deutschen Fledermäuse, Hanau p. 14, 33
Myotis blythii (Tomes, 1857). Proc. Zool. Soc. 1857:53
Myotis punicus Felten, 1977. Senckenb. Biol. 58:1–44
Miniopterus schreibersii (Kuhl, 1817). Die Deutschen Fledermäuse, Hanau p. 14
Carnivora
Felis silvestris Schreber, 1777. Die Säugethiere 3(23):397
Lynx lynx (Linnaeus, 1758). Syst. Nat., 10th ed. 1:43
Lynx pardinus Temminck, 1827. Monogr. Mamm. 1:116
Genetta genetta (Linnaeus, 1758). Syst. Nat., 10th ed. 1:45
Herpestes ichneumon (Linnaeus, 1758). Syst. Nat., 10th ed. 1:43
Urva auropunctatus (Hodgson, 1836). J. Asiat. Soc. Bengal 5:235
Canis aureus Linnaeus, 1758. Syst. Nat., 10th ed. 1:40
Canis lupus Linnaeus, 1758. Syst. Nat., 10th ed. 1:39
Nyctereutes procyonoides (Gray, 1834). Illustr. Indian Zool. 2: pl.1
Vulpes lagopus (Linnaeus, 1758). Syst. Nat., 10th ed. 1:40
Vulpes vulpes (Linnaeus, 1758). Syst. Nat., 10th ed. 1:40
Ursus arctos Linnaeus, 1758. Syst. Nat., 10th ed. 1:47
Ursus maritimus Phipps, 1774. Voyage Towards North Pole, p. 185
Odobenus rosmarus Linnaeus, 1758). Syst. Nat., 10th ed. 1:38
Cystophora cristata (Erxleben, 1777). Syst. Regni Anim. 1:590
Erignathus barbatus (Erxleben, 1777). Syst. Regni Anim. 1:590
Halichoerus gryphus (Fabricius, 1791). Skr. Nat. Selsk. Copenhagen 1(2):167
Monachus monachus (Hermann, 1779). Beschaft. Berlin Ges. Naturforsch. Fr. 4: 501, pls.12,13
Pagophilus groenlandicus (Erxleben, 1777). Syst. Regni Anim. 1:588
Phoca vitulina Linnaeus, 1758. Syst. Nat., 10th ed. 1:38
Pusa hispida (Schreber, 1775). Die Säugethiere 2(13): pl.86 (1775), text 3(17):312 (1776)
Pusa caspica (Gmelin, 1788). Syst. Nat., 13th ed. 1:64
Lutra lutra (Linnaeus, 1758). Syst. Nat., 10th ed. 1:45
Gulo gulo (Linnaeus, 1758). Syst. Nat., 10th ed. 1:45
Martes foina (Erxleben, 1777). Syst. Regni Anim. 1:458
Martes martes (Linnaeus, 1758). Syst. Nat., 10th ed. 1:46
Meles meles (Linnaeus, 1758). Syst. Nat., 10th ed. 1:48
Mustela erminea Linnaeus, 1758. Syst. Nat., 10th ed. 1:46
Mustela eversmannii Lesson, 1827. Manuel de Mammalogie p. 144
Mustela lutreola (Linnaeus, 1761). Fauna Suecica, 2nd ed., p. 5
Mustela nivalis Linnaeus, 1766. Syst. Nat., 12th ed. 1:69
Mustela putorius Linnaeus, 1758. Syst. Nat., 10th ed. 1:46
Neovison vison (Schreber, 1777). Die Säugethiere 3(19): pl.127B (1777), text 3(26):463 (1777)
Vormela peregusna (Güldenstädt, 1770). Nova Comm. Imp. Acad. Sci. Petropoli 14(1):441
Procyon lotor (Linnaeus, 1758). Syst. Nat., 10th ed. 1:48

(continued)

Table 1 (continued)

Nasua nasua (Linnaeus, 1766). Syst. Nat., 12th ed. 1:64
Terrestrial Cetartiodactyla
Sus scrofa Linnaeus, 1758. Syst. Nat., 10th ed. 1:49
Alces alces (Linnaeus, 1758). Syst. Nat., 10th ed. 1:66
Capreolus capreolus (Linnaeus, 1758). Syst. Nat., 10th ed. 1:68
Capreolus pygargus (Pallas, 1771). Reise Prov. Russ. Reichs 1:453
Odocoileus virginianus (Zimmermann, 1780). Geogra. Gesch. Mensch. Vierf. Thiere 2:129
Rangifer tarandus (Linnaeus, 1758). Syst. Nat., 10th ed. 1:67
Axis axis (Erxleben, 1777). Syst. Regn. Anim. 1:312
Cervus elaphus Linnaeus, 1758. Syst. Nat., 10th ed. 1:67
Cervus nippon Temminck, 1838. In von Siebold, Temminck & Schlegel, Fauna Japonica, Coup d'Oeil Faune Iles Sonde Emp. Japan, p. xxii
Dama dama (Linnaeus, 1758). Syst. Nat., 10th ed. 1:67
Hydropotes inermis Swinhoe, 1870. Athenaeum 2208:264
Muntiacus reevesi (Ogilby, 1839). Proc. zool. Soc. 1838:105
Bison bonasus (Linnaeus, 1758). Syst. Nat., 10th ed. 1:71
Capra aegagrus Erxleben, 1777. Syst. Regn. Anim. 1:260
Capra hircus Linnaeus, 1758. Syst. Nat., 10th ed. 1:68
Capra ibex Linnaeus, 1758. Syst. Nat., 10th ed. 1:68
Capra pyrenaica Schinz, 1838. N. Denkschr. Schweiz. Ges. Natur. Wiss. 2:9
Capra caucasica Güldenstaedt et Pallas, 1783. Acta Acad. Sci. Petropoli, for 1779, 2:275
Ovibos moschatus (Zimmermann, 1780). Geogr. Gesch. Mensch. Vierf. Thiere 2:86
Ovis gmelini Blyth, 1841. Proc. Zool. Soc., 1840:69
Rupicapra carpatica Coutourier, 1938. Le Chamois: 369
Rupicapra ornata Neumann, 1899. Ann. Mus. Stor. Nat. Genova 20:347
Rupicapra parva Cabrera, 1911. Proc. Zool. Soc. 1910:999
Rupicapra pyrenaica Bonaparte, 1845. Cat. Meth. Mamm. Europe p. 17
Rupicapra rupicapra (Linnaeus, 1758). Syst. Nat., 10th ed. 1:68
Ammotragus lervia (Pallas, 1777). Spicilegia Zool. 12:12
Cetacea
Balaena mysticetus Linnaeus, 1758. Syst. Nat., 10th ed. 1:75
Eubalaena glacialis (Müller, 1776). Zool. Danicae Prodr. p. 7
Balaenoptera acutorostrata Lacépède, 1804. Hist. Nat. Cetacees p. 134
Balaenoptera bonaerensis Burmeister, 1867. Actas Soc. Paleo., Buenos Aires: 24
Balaenoptera borealis Lesson, 1828. Hist. Nat. Gen. Part. Mamm. Oiseaux 1:342
Balaenoptera edeni Anderson, 1879. Anat. Zool. Res., Yunnan:551, pl.44
Balaenoptera musculus (Linnaeus, 1758). Syst. Nat., 10th ed. 1:76
Balaenoptera physalus (Linnaeus, 1758). Syst. Nat., 10th ed. 1:75
Megaptera novaeangliae (Borowski, 1781). Gemein. Naturgesch. Thier. 2(1):21
Eschrichtius robustus (Lilljeborg, 1861). Forh. Skand. Naturf. Ottende Mode, Kopenhagen 1860, 8:602 (1861)
Delphinus delphis Linnaeus, 1758. Syst. Nat., 10th ed. 1:77
Feresa attenuata Gray, 1874. Ann. Mag. Nat. Hist. ser. 4, 14:238–239
Peponocephala electra Gray, 1846. Zoology of the voyage of H.M.S. Erebus and Terror, 1(Mammalia):48, pl.1, fig. 1, p. 35, pl.13
Globicephala macrorhynchus Gray, 1846. Zool. Voy. H.M.S. "Erebis" and "Terror" 1:33
Globicephala melas (Traill, 1809). Nicholson's J. Nat. Philos. Chem. Arts 22:81
Grampus griseus (G. Cuvier, 1812). Ann. Mus. Hist. Nat. Paris 19:13
Lagenodelphis hosei Fraser, 1956. Sarawak Mus. J., n.s. 8(7):496
Lagenorhynchus acutus (Gray, 1828). Spicil. Zool. 1:2

(continued)

Table 1 (continued)

Lagenorhynchus albirostris (Gray, 1846). Ann. Mag. Nat. Hist. ser. 1, 17:84
Orcinus orca (Linnaeus, 1758). Syst. Nat., 10th ed. 1:77
Pseudorca crassidens (Owen, 1846). Hist. Brit. Foss. Mamm. Birds p. 516, fig. 213
Sousa plumbea (G. Cuvier, 1829). Règne Anim. 1:288
Stenella clymene Gray, 1846. Zool. Voy. H.M.S. "Erebis" and "Terror" 1:39
Stenella coeruleoalba (Meyen, 1833). Nova Acta Acad. Caes. Nat. Curios. 16(2):609, pl.43
Stenella frontalis (G. Cuvier, 1829). Règne Anim. 1:288
Stenella longirostris (Gray, 1828). Spicil. Zool. 1:1
Steno bredanensis (G. Cuvier in Lesson, 1828). Hist. Nat. Gen. Part. Mamm. Oiseaux 1:206
Tursiops truncatus (Montagu, 1821). Mem. Wernerian Nat. Hist. Soc. 3:75, pl.3
Delphinapterus leucas (Pallas, 1776). Reise Prov. Russ. Reichs 3(1):85 (footnote)
Monodon monoceros Linnaeus, 1758. Syst. Nat., 10th ed. 1:75
Phocoena phocoena (Linnaeus, 1758). Syst. Nat., 10th ed. 1:77
Kogia breviceps (Blainville, 1838). Ann. fr. étrang. Anat. Physiol. 2:335–337
Kogia sima (Owen, 1866). Trans. Zool. Soc. London 6(1):30, pls.10–14
Physeter macrocephalus Linnaeus, 1758. Syst. Nat., 10th ed. 1:76
Hyperoodon ampullatus (Forster, 1770). In Kalm, Travels into N. Amer. 1:18
Mesoplodon bidens (Sowerby, 1804). Trans. Linn. Soc. Lond. 7:310
Mesoplodon densirostris (Blainville, 1817). Nouv. Dict. Hist. Nat., Nouv. Ed. 9:178
Mesoplodon europaeus (Gervais, 1855). Hist. Nat. Mammifères 2:320
Mesoplodon grayi Von Haast, 1876. Proc. Zool.Soc. Lond. 1876:9
Mesoplodon mirus True, 1913. Smithsonian Miscell. Collection 60(25):1
Ziphius cavirostris G. Cuvier, 1823. Rech. Oss. Foss., Nouv. Ed. 5(1):350
Rodentia
Sciurus anomalus Gmelin, 1778. In Linnaeus, Syst. Nat., 13th ed. 1:148
Sciurus carolinensis Gmelin, 1778. In Linnaeus, Syst. Nat., 13th ed. 1:148
Sciurus vulgaris Linnaeus, 1758. Syst. Nat., 10th ed. 1:63
Callosciurus erythraeus (Pallas, 1779). Nov. Sp. Quad. Gli. Ord.:377
Callosciurus finlaysonii (Horsefield, 1823). Zool. Res. Java 7:151
Atlantoxerus getulus (Linnaeus, 1758). Syst. Nat., 10th ed. 1:64
Marmota bobak (Müller, 1776). Linné's Vollständ. Natursyst. Suppl. p. 40
Marmota marmota (Linnaeus, 1758). Syst. Nat., 10th ed. 1:60
Spermophilus citellus (Linnaeus, 1766). Syst. Nat., 12th ed. 1:80
Spermophilus fulvus (Lichtenstein, 1823). In Eversmann, Reise von Orenburg nach Buchara p. 119
Spermophilus major (Pallas, 1779). Nova Spec. Quad. Glir. Ord. p. 125
Spermophilus pygmaeus (Pallas, 1778). Nova Spec. Quad. Glir. Ord. p. 122
Spermophilus suslicus (Güldenstaedt, 1770). Nova Comm. Acad. Sci. Petropoli 14:389
Eutamias sibiricus (Laxmann, 1769). Sibirische Briefe, Göttingen p. 69
Tamias striatus (Linnaeus, 1758). Syst. Nat., 10th ed. 1:64
Dryomys nitedula (Pallas, 1778). Nova Spec. Quad. Glir. Ord. p. 88
Eliomys quercinus (Linnaeus, 1766). Syst. Nat., 12th ed. 1:84
Muscardinus avellanarius (Linnaeus, 1758). Syst. Nat., 10th ed. 1:62
Myomimus roachi (Bate, 1937). Ann. Mag. Nat. Hist. ser. 10, 20:399
Glis glis (Linnaeus, 1766). Syst. Nat., 12th ed. 1(1):87
Castor canadensis Kuhl, 1820. Beitr. Zool. Vergl. Anat. 1:64
Castor fiber Linnaeus, 1758. Syst. Nat., 10th ed. 1:58
Allactaga major (Kerr, 1792). Anim. Kingd.:274
Dipus sagitta (Pallas, 1773). Reise 2:706
Stylodipus telum (Lichtenstein, 1823). In Eversmanns Reise:120

(continued)

Table 1 (continued)

Pygeretmus pumilio (Kerr, 1792). Anim. Kingd.:275
Scarturus elater (Lichtenstein, 1828). Abh. k. Akad. Wiss. Berlin:155
Sicista betulina (Pallas, 1779). Nova Spec. Quadr. Glir. Ord. p. 332
Sicista caucasica Vinogradov, 1925. Proc. zool. Soc.:548
Sicista kazbegica Sokolov, Baskevich & Kovalskaya, 1986. Zool. Zh., 65(6):949
Sicista kluchorica Sokolov, Kovalskaya & Baskevich, 1980. Gryzuny Severnovo Kavkaza:38
Sicista nordmanni (Keyserling and Blasius, 1840). Wirbelth. Europas:38
Sicista severzovi Ognev, 1935. Byulletin Nauchno-issled. Inst. Zool. Mosk. 2: 54
Sicista strandi (Formozov, 1931). Folia Zool. Hydrob. Riga 3:79.
Sicista subtilis (Pallas, 1773). Reise Prov. Russ. Reichs. 1(2):705
Sicista trizona (Frivaldszky, 1865). Termeszetrajzi Ftizetek, 5:103
Nannospalax leucodon (Nordmann, 1840). Demidoff Voy. 3:34
Nannospalax xanthodon (Nordmann, 1840). Demidoff Voy. 3:35
Spalax antiquus Mehely, 1909. A Földi Kutyák Fajai Budap.:175
Spalax arenarius Reshetnik, 1939. Reports Zool. Mus. Kiev 23:11
Spalax giganteus Nehring, 1898. Sitzber. Ges. Naturf. Frde Berlin p. 169
Spalax graecus Nehring, 1898. Zool. Anz. 21:479–481
Spalax istricus Mehely, 1909. A Földi Kutyák Fajai Budap.:186
Spalax microphthalmus Güldenstaedt, 1770. Nova Comm. Acad. Sci. Petropoli 14:1
Spalax zemni (Erxleben, 1777). Syst. Regni Anim. 1:370–371
Prometheomys schaposchnikowi Satunin, 1901. Zool. Anz. 24:574
Arvicola amphibius (Linnaeus, 1758). Syst. Nat., 10th ed. 1:61
Arvicola sapidus Miller, 1908. Ann. Mag. Nat. Hist. ser. 8, 1:195
Arvicola italicus Savi, 1839. Nuovo Giorn. de Lett., Pisa 37, 102:202
Chionomys gud (Satunin, 1909). Izv. Kaukas. Mus. 4:272
Chionomys nivalis (Martins, 1842). Rev. Zool. Paris p. 331
Chionomys roberti (Thomas, 1906). Ann. Mag. Nat. Hist. 17:418
Craseomys rufocanus (Sundevall, 1846). Ofv. K. Svenska Vet.-Akad. Forhandl. Stockholm 3:122
Dicrostonyx torquatus (Pallas, 1778). Nov. Spec. Quad. Gli. Ord.:77
Dinaromys bogdanovi (V. Martino & E. Martino, 1922). Ann. Mag. Nat. Hist. ser. 9, 9:413
Ellobius talpinus (Pallas, 1770). Nova Comm. Acad. Sci. Petropoli 14, 1:568
Lagurus lagurus (Pallas, 1773). Reise Prov. Russ. Reichs. 2:704
Lemmus lemmus (Linnaeus, 1758). Syst. Nat., 10th ed. 1:59
Lemmus sibiricus (Kerr, 1792). Anim. Kingd.:241
Alexandromys middendorfii (Poljakov, 1881). Mem. Imp. Acad. Sci. St. Petersb. 39: appendix 2:70
Alexandromys oeconomus (Pallas, 1776). Reise Prov. Russ. Reichs. 3:693
Microtus agrestis (Linnaeus, 1761). Fauna Suecica, 2nd ed. p. 11
Microtus levernedii (Crespon, 1844). Faune Meridionale, 1:73
Microtus rozanius (Bocage, 1865). Mem. Ac. Real. Sci. de Lisboa, 3,2:7
Microtus arvalis (Pallas, 1778). Nova Spec. Quadr. Glir. Ord. p. 78
Microtus brachycercus (Lehmann, 1961). Zool. Anz. 167:223
Microtus cabrerae (Thomas, 1906). Ann. Mag. Nat. Hist. ser. 7, 17:576
Microtus daghestanicus (Shidlovsky, 1919). Tiflis Bull. Terr. Exper. Stat. 2:22
Microtus duodecimcostatus (Selys-Longchamps, 1839). Rev. Zool. Paris p. 8
Microtus felteni Malec & Storch, 1963. Senckenbergiana biol. 44:171
Microtus gerbei (Gerbe, 1879). Le Naturaliste 1:51
Microtus hartingi (Barrett-Hamilton, 1903). Ann. Mag. Nat. Hist. 11:307
Microtus levis Miller, 1908. Ann. Mag. Nat. Hist. ser. 8, 1:197
Microtus liechtensteini Wettstein, 1927. Anz. Akad. Wiss., Wien 20:2
Microtus lusitanicus (Gerbe, 1879). Rev. Mag. Zool. Paris ser. 3, 7:44

(continued)

Table 1 (continued)

Microtus majori Thomas, 1906. Ann. Mag. Nat. Hist., ser. 7, 17:419
Microtus multiplex (Fatio, 1905). Arch. Sci. Phys. Nat. Geneve, ser. 4, 19:193
Microtus nebrodensis (Mina-Palumbo, 1868). Ann. Agric. Sicil. 12:61
Microtus savii (de Selys-Longchamps, 1838). Rev. Zool. Paris p. 248
Microtus socialis (Pallas, 1773). Reise Prov. Russ. Reichs. 2:705
Microtus subterraneus (de Selys-Longchamps, 1836). Essai Monogr. sur les Campagnols des Env. de Liege, p. 10
Microtus tatricus (Kratochvíl, 1952). Acta Acad. Sci. Nat. Moravo-Siles. 24:155–194
Microtus thomasi (Barrett-Hamilton, 1903). Ann. Mag. Nat. Hist., Ser. 7, 11:306
Stenocranius gregalis (Pallas, 1779). Nov. Spec. Quad. Gli. Ord.:238
Clethrionomys glareolus (Schreber, 1780). Die Säugethiere 4: 680. See Kryštufek et al. (2019)
Clethrionomys rutilus (Pallas, 1779). Nova Spec. Quadr. Glir. Ord., p. 246
Myopus schisticolor (Lilljeborg, 1844). Ofv. K. Svenska Vet.-Akad. Forhandl. Stockholm I:33
Ondatra zibethicus (Linnaeus, 1766). Syst. Nat., 12th ed. 1:79
Allocricetulus eversmanni (Brandt, 1859). Mel. Biol. Acad. St. Pétersb.:210
Cricetus cricetus (Linnaeus, 1758). Syst. Nat., 10th ed. 1:60
Mesocricetus brandti (Nehring, 1898). Zool.Anz. 21:331
Mesocricetus newtoni (Nehring, 1898). Zool. Anz. 21:329
Notocricetulus migratorius (Pallas, 1773). Reise Prov. Russ. Reichs. 2:703
Acomys cahirinus (É. Geoffroy, 1803). Cat. Mam. Mus. Natl. Hist. Nat., Paris:195
Apodemus agrarius (Pallas, 1771). Reise Prov. Russ. Reichs. 1:454
Apodemus alpicola Heinrich, 1952. J. Mammal. 33:260
Apodemus epimelas (Nehring, 1902). Sitz. Ber. Ges. Naturf. Fr. Berlin 1902:2
Apodemus flavicollis (Melchior, 1834). Dansk. Staat. Norg. Pattedyr, p. 99
Apodemus mystacinus (Danford & Alston, 1877). Proc. Zool. Soc. Lond. 1877:279
Apodemus sylvaticus (Linnaeus, 1758). Syst. Nat., 10th ed. 1:62
Apodemus uralensis (Pallas, 1811). Zoogr. Rosso-Asiat. 1:168
Apodemus witherbyi (Thomas, 1902). Ann. Mag. Nat. Hist., ser. 7, 10:490
Micromys minutus (Pallas, 1771). Reise Prov. Russ. Reichs. 1:454
Mus cypriacus Cucchi, Orth, Auffray, Renaud, Fabre, Catalan, Hadjisterkotis, Bonhomme, Vigne, 2006. Zootaxa 1241:1–36
Mus macedonicus Petrov & Ruzic, 1983. Proc. Fauna SR Serbia, Serbian Acad. Sci. and Arts, Belgrade 2:177
Mus musculus Linnaeus, 1758. Syst. Nat., 10th ed. 1:62
Mus spicilegus Petényi, 1882. Termeszetrajzi Fuzetek, Budapest 5:114
Mus spretus Lataste, 1883. Actes Soc. Linn. de Bordeaux, ser. 7, 4:27
Rattus norvegicus (Berkenhout, 1769). Outlines Nat. Hist. Great Britain and Ireland, 1:5
Rattus rattus (Linnaeus, 1758). Syst. Nat., 10th ed., 1:61
Meriones meridianus (Pallas, 1773). Reise Russ. Reichs 2:702
Meriones tamariscinus (Pallas, 1773). Reise Russ. Reichs 2:702
Meriones tristrami Thomas, 1892. Ann. Mag. Nat. Hist. 9:148
Hystrix cristata (Linnaeus, 1758). Syst. Nat., 10th ed., 1:56
Myocastor coypus (Molina, 1782). Sagg. Stor. Nat. Chile, p. 287
Diprotodontia
Macropus rufogriseus (Desmarest, 1817). Nouv. Dict. Hist. Nat., Nouv. Ed. 17:36

(e.g., Trouessart 1910) and English (e.g., Barrett-Hamilton 1913). At about that time (1919), the American Society of Mammalogists was founded along with the quarterly *Journal of Mammalogy* (Hoffmeister and Sterling 1994).

The first professional mammal society in Europe was founded in 1926 as Deutsche Gesellschaft für Säugetierkunde (German Society for Mammalian Biology) with its journal *Zeitschrift für Säugetierkunde* (since 2001 *Mammalian Biology*).

Mammals have been translated to German as Säugetiere (literary sucking animals) and the science devoted to their study received a name Säugetierkunde (Kunde is German for science). In 1936 the Natural History Museum in Paris started publishing the periodical *Mammalia* (now published by de Gruyter). In 1954 the Mammal Society was founded in the UK, with a periodical *Mammal Review* starting in 1970. Mammalogy has been avoided in all these attempts. The first professional mammal society in Europe to adopt the name mammalogy was seemingly the Mammalogical Section of the Natural History Society at the National Museum in Prague in 1958 (In Czech: *Mammaliologické sekce Přírodovědeckého sboru Společnosti Národního Musea*), together with the periodical of a similarly long name in 1959 (*Mammaliologické zprávy/Novitas mammaliologicae. Nová série/Series nova,* 1962 renamed as *Lynx (n.s.) Praha*). The name "mammaliologické" was difficult for pronunciation and several European languages gave priority to theriology. Czechs and Italians, for example, interchangeably used in the past, and still now both, mammalogy and theriology. Thus, the Italian Mammal Society (its official name in English) is called in Italian Associazione Teriologica Italiana and is issuing a periodical with the English title *Hystrix: Italian Journal of Mammalogy* since 1986.

The term theriology was more widely accepted in eastern Europe which was evident from the periodical *Acta Theriologica* (since 2015 *Mammal Research*) issued since 1954 by the Mammal Research Institute of the Polish Academy of Sciences (founded in 1952) and from the professional mammal society founded in the former Soviet Union under the name All Union Theriological Society. The Society involved 12 regional sections and after the collapse of the Soviet Union some of these sections continued their work as independent societies, for example, Russian Theriological Society (since 1992; the official name is Russian Theriological Society of the Russian Academy of Science), Ukrainian Theriological Society (1993), and Lithuanian Theriological Society (1989). The term mammalogy is not in use in these countries, and also the journals published by the societies avoid it: *Theriologia Ukrainica* (started in 1998 as *Proceedings of the Theriological School*), *Russian Journal of Theriology* (founded in 2002), and *Theriologia Lituanica* (not published continuously).

Other 27 mammalogical journals founded in Europe after the Second World War are listed in Table 2. There are more local leaflets or journals on bats and other organisms with a more local distribution.

Taxonomy

The discovery of about 320 indigenous and introduced mammal species (Table 1) in Europe took more than 250 years. We list 322 species, but the number of acknowledged species will differ depending on the underlying taxonomic philosophy and species concept(s). As already mentioned, the formal system of nomenclature was developed by C. Linnaeus. His basic works (Linnaeus 1758, 1761, 1766, 1768) listed and named 77 species of mammals from Europe, most of which are currently regarded as valid species. Further new European species of mammals were described by Berkenhout (1769), Laxmann (1769, 1788), Forster (1770), Gmelin (1770, 1778, 1788), Güldenstaedt (1770), Pallas (1769, 1770, 1771, 1773, 1775, 1776, 1777, 1778, 1779, 1811), Pennant (1771), Schreber (1774, 1775, 1777, 1780), Phipps (1774), Müller (1776), Erxleben (1777), Hermann (1779, 1780), Zimmermann (1780), Borowski (1781), Molina (1782), Güldenstaedt and Pallas (1783), Fabricius (1791), Kerr (1792), Borkhausen (1797), Bechstein (1800), Shaw (1801), Lacépède (1804), Sowerby (1804), E. Geoffroy (1803, 1806, 1810, 1811, 1818), Traill (1809), Cuvier (1812, 1823, 1829), Rafinesque (1814), Desmarest (1817), Blainville (1817, 1838, 1839), Kuhl (1817, 1820), Montagu (1821), Savi (1822, 1839), Horsefield (1823), Liechtenstein (1823, 1828), Boie (1825), Leach (1825), Gray (1812, 1828, 1834, 1846, 1874), Lesson (1827, 1828), Temminck (1827, 1838, 1840), Cuvier in Lesson (1828), Millet (1828), Fischer (1829), Meyen (1833), Melchior (1834), Küster (1835), Hodgson (1836), Sviridenko

Table 2 An overview of mammal journals published by European societies or institutions

Year	Journal name
1926–2000	*Zeitschrift für Säugetierkunde*, continued as (2001–present) *Mammalian Biology*
1936–present	*Mammalia*
1952–present	*Lutra*
1952–2001	*Säugetierkundliche Mitteilungen*
1954–2014	*Acta Theriologica*, continued as (2015–present) *Mammal Research*
1958–1961	*Mammaliologické zprávy/Novitas mammaliologicae. Nová série/Series nova*, continued as (1962–present) *Lynx, n.s.* (Praha)
1960–present	*Folia Primatologica*
+1961–1999	*Equus*
1963–2005	*Myotis*
1969–present	*Nyctalus*
1969–1994	*Investigations on Cetacea*
1970–present	*Mammal Review*
1970–2001	*Säugetierschutz*
1977–present	*Säugetierkundliche Informationen*
1978–2003?	*Przewalski Horse*
1981–1998?	*Eliomys* (Gent)
1984–2010?	*Arvicola*
1985–present	*Le Rhinolophe*
1986–present	*Hystrix: Italian Journal of Mammalogy*
1987–present	*Der Flattermann*
1988–present	*Galemys, Spanish Journal of Mammalogy*
1990–present	*Zoogdier*
1991–present	*MAUS*, Mitteilungen aus unserer Säugetierwelt
1993–present	*Journal of Mammalian Evolution*
1996–2000	*Folia Theriologica Estonica*
1996–2016	*Vespertilio*
1998–present	*Plecotus* et al.
1998–present	*Proceedings of the Theriological School*, continued as *Theriologia Ukrainica*
1999–present	*Acta Chiropterologica*
1999–present	*Mitteilungen für sächsische Säugetierfreunde*
2000–2002	*pro chiroptera*
2000–2007	*Studia Chiropterologica*
2000–2009	*Nietoperze*
2002–present	*Theriologia Lituanica*
2002–present	*Russian Journal of Theriology*

(1936), Sélys-Longchamps (1836, 1838, 1839), Bonaparte (1837, 1840, 1845), Martin (1838), Schinz (1837, 1838), Ogilby (1839), Keyserling and Blasius (1839, 1840), Lereboullet (1842), Nordmann (1840), Blyth (1841), Martins (1842), Lilljeborg (1844), Crespon (1844), Eversmann (1845), Owen (1846, 1866), Sundevall (1846), Blasius (1853), Gervais (1855), Rosenhauer (1856), Brandt (1859), Tomes (1857), Lilljeborg (1861), Bocage (1865), Frivaldszky (1865), Burmeister (1867), Peters (1867, 1869), Mina-Palumbo (1868), Swinhoe (1870), Von Haast (1876), Danford and Alston (1877, 1880), Anderson (1878), Dobson (1878), Gerbe (1879), Poljakov (1881), Petényi (1882), Lataste (1883), Monticelli (1885), Satunin (1895, 1901), Allen (1890), De Winton (1898), Nehring (1894, 1898, 1902), Neumann (1899), Merriam (1900), Miller (1900, 1908, 1910), Thomas (1892, 1901, 1902, 1906), Barrett-Hamilton (1900, 1903, 1907), Matschie (1901), Cabrera (1904, 1907, 1911), Fatio (1905), Bate (1906, 1937), Barrett-Hamilton (1907), Mottaz (1907), Miller (1908, 1910), Satunin (1908, 1909), Mehely (1909), True (1913), Shidlovsky (1919), Martino, V. &. E. (1922), Ognev (1922, 1924, 1935), Turov (1924), Vinogradov (1925), Altobello (1926), Formozov (1931), Martino, V. &. E. (1931), Kormos (1934), Kuzjakin (1935, 1965), Bate (1937), Zubko (1937), Coutourier (1938), Reshetnik (1939), Heinrich (1952), Kratochvíl (1952), Kratochvíl and Rosicky (1952), Wettstein (1927, 1953), Fraser (1956), Lehmann (1961, 1964), König (1962), Malec and Storch (1963), Ondrias (1966), Palacios (1977), Felten (1971, 1977), Djulic (1980), Sokolov, Kowalskaya and Baskevich (1980), Petrov and Ružić (1983), Sokolov, Baskevich and Kowalskaya (1986), Hutterer, López-Jurado, and Vogel (1987), de Paz (1994), Helversen and Heller (2001), Mucedda et al. (2002), Benda et al. (2004), Hulva and Benda (2004), Cucchi et al. (2006), Nicolas, Martínez-Vargas, and Hugot (2017), Kryštufek et al. (2018) and Ruedi et al. (2019). References are given in Table 2. Subspecific names, a possible source for further species names, are not listed here. More species will be recognized after biogeographical studies have been finished, such as for voles (e.g., Jaarola and Searle 2002), or shrews (Amori and Castiglia 2018). See also Burgin et al. (2018) for a discussion of a recent species list and Genovesi et al. (2009) for a review of alien species.

In parallel with the discoveries of new species, mammalogists of the late eighteenth and nineteenth centuries also built up regional lists of species. Progress in cataloguing the mammal richness was not a steady accumulation of knowledge, but rather a series of ups and downs. The nineteenth-century European mammalogy reached its pinnacle in mid-century (1857) in The Natural History of Mammals of Germany and adjacent regions of Central Europe by the German Johann Heinrich Blasius (1809–1870). His work remained in high esteem for the rest of the century on the one hand, but also created an illusion that not much new could be expected in Europe on the other hand. European mammalogists, confronted with the European mammal fauna, which seemed not to be particularly challenging, and the challenges offered by overseas colonial possessions, chose the latter.

In the meantime, mammalogy in the New World progressed rapidly both conceptually and methodologically. Cuvier's concept of immutable species and varieties was replaced by polytypic species and subspecies. Study of variation emerged as the central topic in mammalogy which demanded clearer and more intelligible diagnostics of taxa. This could not be achieved without detailed descriptions of cranial and dental morphology, in addition to external appearance, and meticulous morphometrics for quantifying size and proportions. Above all, comparisons between taxa necessitated samples (hypodigms), not just individuals, and such demands could no longer be satisfied by taxidermic mounts. A whole series of conspecific individuals had to be sampled in the field, measured, prepared in a standardized way, and deposited in museum collections for further study. A museum voucher became a standard in taxonomic work. It consisted of a skin and skull with attached label containing detailed information on the locality, date of collecting, sex, standard external measurements, and relevant details on the habitat and observations made during dissection. The mammal collections were still small and scrappy in the 1880s and 1890s, and small mammals in particular were heavily underrepresented. At that time, Clinton Hart Merriam (1855–1942) at the United States Department of Agriculture (founded in 1885 and renamed in 1905 to Bureau of Biological Survey) started collecting small mammals using commercially available traps called the "Cyclone." The trap "was an affair of tin and wire springs, only about two inches square when collapsed, cheap in cost, and easily portable in quantity" (Osgood 1944). Simultaneously, the American mammalogists modified the way of skinning birds as museum vouchers, developed earlier on by ornithologists. Application of these two novelties enlarged mammal collection in the US Museums to proportions which at that time were unprecedented. Still in 1910, Edouard-Louis Trouessart (1842–1927) from the Natural History Museum in Paris wrote with amazement of the US collections in which common species were represented by series numbering up to 1200 museum vouchers (Denys et al. 2012).

In the 1890s Gerrit Smith Miller (1869–1956), at that time still employed at the Department of Agriculture (in 1898 he moved to the United States National Museum) transplanted to Europe "the methods and philosophy of the 'American School of Mammalogy', including the systematic study of large series of uniformly prepared small mammal specimens" (Dunnum and Cook 2012). In 1894, when Miller demonstrated at the British Museum new methods of field collecting small mammals using the "Cyclone" traps and processing the material as standard museum vouchers, European mammalogists were deeply impressed. The method was quickly adopted in various European countries, for example, France, Germany, and Russia.

European Mammal Collections

Europe has a long history of biological collections. Collections for scientific purposes (Genoways and Schlitter 1981) are younger and often are subject to change. Only larger collections run by public institutions have a chance to survive for longer periods. Table 3 lists some current collections where

Table 3 European mammal collections containing about 2,000 or more specimens of recent mammals

Institution	Number of specimens
Natural History Museum, London, UK	333,000
Rijksmuseum van Natuurlijke Historie, Leiden, The Netherlands	300,000
Zoological Museum of Moscow State University, Moscow, Russia	201,000
Mammal Research Institute, PAS, Bialowieza, Poland	190,000
Institute of Plant and Animal Ecology, Ural Branch of the Russian Academy of Sciences, Yekatarinburg, Russia	183,000
Museum für Naturkunde, Berlin, Germany	150,000
Koninklijk Museum vaar Midden-Africa, Tervuren, Belgium	135,000
Muséum National d'Histoire Naturelle, Paris, France	130,000
Naturhistorisches Museum, Vienna, Austria	125,000
Zoologisches Forschungsmuseum Alexander Koenig, Bonn, Germany	120,000
Zoological Institute, St. Petersburg, Russia	100,000
Naturhistoriska Riksmuseet, Stockholm, Sweden	100,000
SNS, Forschungsinstitut Senckenberg, Frankfurt a.M., Germany	95,000
Finnish Museum of Natural History, Helsinki, Finland	60,000
Grant Museum of Zoology, University College London, London, UK	60,000
Staatliches Museum für Naturkunde, Stuttgart, Germany	51,000
Museum d'Historie Naturelle, Geneve, Switzerland	45,950
National Museum of Scotland, Edinburgh, UK	45,000
Hungarian Natural History Museum, Budapest, Hungary	44,276
Institut Royal des Sciences Naturelles de Belgique, Brussels, Belgium	42,000
Zoologische Staatssammlung, München, Germany	40,000
Natural History Museum of Denmark, Copenhagen, Denmark	40,000
Institute of Vertebrate Zoology, CAS, Brno, Czech Republic	40,000
Harrison Institute, Sevenoaks, UK	38,000
National Museum, Praha, Czech Republic	37,500
Department of Zoology, Charles University, Praha, Czech Republic	30,000
Biological Museum, Lund University, Sweden	30,000
Slovenian Museum of Natural History, Ljubljana, Slovenia	28,000
National Museum of Natural Sciences, Madrid, Spain	27,000
SNS, Museum für Tierkunde, Dresden, Germany	25,000
Zoologisches Museum, Hamburg, Germany	23,000
Museo Zoologico de La Specola, Firenze, Italy	22,330
SNS, Senckenberg Museum für Naturkunde, Görlitz, Germany	22,000
Landessammlungen für Naturkunde, Karlsruhe, Germany	21,000
National Museum of Natural Sciences, Kiev, Ukraine	20,300
Zoological Museum, Oulu, Finland (closed)	20,000
Biological Museum, Lund University, Sweden	18,700
Centrum für Naturkunde, Hamburg, Germany	17,000
Natural History Museum, Oslo, Norway	12,000
National Museum of Ireland, Dublin, Ireland	10,500
"Grigore Antipa" Museum of Natural History, Bucharest, Romania	10,000
Naturhistorisches Museum, Bern, Switzerland	10,000
Naturhistorisches Museum, Basel, Switzerland	9,500
Institute of Zoology, Academy of Sciences, Sofia, Bulgaria	8,400
Zoological Museum, University of Odesa, Odesa, Ukraine	8,000
Übersee-Museum Bremen, Germany	7,500
Department of Zoology, World Museum Liverpool, Liverpool, UK	7,000

(continued)

Table 3 (continued)

Institution	Number of specimens
Zoological Museum of the Kiev's State University, Kiev, Ukraine	7,000
Zoologisches Museum der Universitat Zürich, Zürich, Switzerland	7,000
Zoological Museum, University of Uzhgorod, Uzhgorod, Ukraine	6,513
National Museum of Natural History, Lvov, Ukraine	6,476
Laboratorium voor Algemende Dierkunde, Antwerpen, Belgium	6,000
Quex Museum, Birchington, UK	6,000
Museuo Civico di Storia Naturalle di Milano, Milano, Italy	5,700
Manchester Museum, Manchester, UK	5,367
Zoological Institute, Tartu University, Tartu, Estonia	5,344
Oxford University Museum, Oxford, UK	5,000
Zoologie der Sektion Biowissenschaften der Martin-Luther-Universitat, Halle, Germany	4,200
Estonian Museum of Natural History, Tallin, Estonia	4,100
All-Union Research Institute of Game Management and Fur Farming, Kirov, Russia	3,500
Museuo Civico di Storia Naturalle di Verona, Verona, Italy	3,500
Museo e Instituto de Zoologia Sistematica, Universa di Torino, Torino, Italy	3,500
Zoological Museum, University of Lvov, Ukraine	3,247
Zoological Museum, University of Dnepropetrovsk, Dnepropetrovsk, Ukraine	3,156
National Museum of Ireland, Dublin, Ireland	3,000
Museum of Evolution, Upsala, Sweden	2,000

Data obtained in March 2019. Numbers are estimates

mammals are kept for scientific studies and/or for public display. The listed mammal collections sum up to 3.2 million of specimens.

European Mammal Societies

After the First and Second World Wars, some national societies for the study and conservation of mammals were founded. One of the earliest ones was founded in Germany (1926), followed by The Netherlands (1952), France (1954), the United Kingdom (1954), Czechoslovakia (1958), Italy (1983), Lithuania (1989), Russia (1992), Ukraine (1993), and Spain (2000). Numerous local societies for the conservation of bats, dormice, hamsters, hedgehogs, large carnivores, otters, etc., were also founded in various European countries.

Mammal Congresses

Scientific congresses on mammalian topics have been held in Germany by the German Society of Mammalogist almost annually since 1926 (Hutterer 2001), in France by the French Society for the Protection of Mammals (13th Colloque International de Mammalogie in Banyuls, 1989), and certainly also by many other national societies. The European Mammal Foundation has held congresses since 1991 (Lisbon), the eighth one being organized in Warsaw in 2019. In 1960 and 1971, early meetings were held in Brno, Czechoslovakia. The International Theriological Congress was first organized in Moscow in 1974 and has been continued under the name International Mammalogical Congress since 2001 (Lidicker 2011).

Handbooks

Despite all the engagements in overseas explorations, several European countries printed mammal faunas of their territories already in the second half of the nineteenth century: UK (Lydekker 1896, Johnston 1903, Barrett-Hamilton 1910–1921, 1913), Germany (Blasius 1857), or Spain (Graells 1897). There was a need, however, for a comprehensive treatise at the continental scale to standardize taxonomy and nomenclature. In 1910, Trouessart, at that time appointed at the Mammals and Birds section of the National Museum of

Natural History in Paris (Denys et al. 2012), published the "Fauna of the Mammals of Europe" (Trouessart 1910). Shortly afterwards, an even more influential work followed, authored by Miller. Several mammalogists in London, Lord Lilford (Thomas Littleton Powys, 1833–1896), Oldfield Thomas (1858–1929), and Gerald Edwin Hamilton Barrett-Hamilton (1871–1914) put huge efforts in completing collections of European mammals which around 1910 contained 5000 museum vouchers, including 124 types. This material, along with 4000 vouchers held in Washington, and further 2500 museum specimens scattered across Europe, allowed Miller to produce a monographic treatise, a Catalogue of mammals of "Europe exclusive of Russia" on more than one thousand pages (Miller 1912). Miller recognized 314 "forms" (species and subspecies) in 69 genera. He himself examined museum vouchers of all these forms except six. Miller's Catalogue was much more than just a list of species. It contained detailed morphological descriptions, accompanied by craniodental measurements and drawings of skulls and dentition, produced by Amedeo John Engel Terzi (1872–1956). The quality of illustrations is such that they are still reproduced in textbooks of mammalogy. Furthermore, the Catalogue included dichotomous keys to families, genera, species and subspecies, and lists of all vouchers examined, together with localities and other details like sex, date, and collector. The Catalogue was a model for the most important contributions to European mammalogy (see Shamel et al. 1954) and served as the taxonomic standard well into the 1970s and 1980s. Even today, the Catalogue remains to be an invaluable nomenclatural source and a reference for morphological data. Contrary to Miller, Trouessart also considered marine mammals and covered Europe in its entirety, that is, as far as the Urals in the east and the Caucasus in the southeast. Despite the broader geographic and taxonomic scope, Trouessart's book counts only 266 pages (as compared to 1019 pages in Miller's Catalogue) which was a consequence of less detailed descriptions and lack of illustrations and identification keys. Miller's Catalogue prevailed because it allowed mammalogists a more secure classification of their vouchers and recognition of still unnamed taxa.

Miller in his Catalogue strictly adhered to Europe west of Russia, an evident consequence of the paucity of material from the East, both in general, and in particular in major museums of Central and West Europe. At about same time, the Russian mammologist Sergey I. Ognev (1886–1951) published "Fauna Mosquensis" which was supported by about 3000 museum vouchers (Bakloushinskaya et al. 2012). Despite such parallel trends in the West and the East, Miller's geographic scope proved remarkably persistent, being uniformly followed by subsequent authors well into the 1980s. This was not a matter of free choice, but of political reality in Europe during the twentieth century. In the same year that Miller published his Catalogue, a local conflict erupted in the Balkans, mammalogically the least known region in Europe. The skirmish soon became known as the First Balkan War. It was followed in 1913 by a brief Second Balkan War and in 1914 by the Third Balkan War which escaped control and developed into World War I. When the Great War, as it was called at the time, ended, the political map of Europe had been redrawn. The continent was instable, insecure, impoverished, and ideologically divided. Miller's border became a reality and West and East Europe took their own courses in mammalogy with not much collaboration. The environment therefore did not encourage mammal research and not many syntheses on mammals were published on either side of the border between the two wars.

Mammals were treated in handbooks at different levels. Mammals of the Palaearctic Region were covered by Ellerman and Morrison-Scott (1957, 1966) and Corbet (1978, 1984), and European mammals by Keyserling and Blasius (1840), Schmiedeknecht (1906), Trouessart (1910), Miller (1912), Hainard (1948, 1949), van den Brink (1955), Gaffrey (1961), Corbet (1966), König (1969), Curry-Lindahl (1975), Corbet and Ovenden (1980), Schilling et al. (1983), Bjärvall and Ullström (1986), Görner and Hackethal (1988), and Lange et al. (1994). Niethammer and Krapp (1978–2005) presented the first detailed handbook series, and Macdonald and Barrett (1993) provided an overview of British and European mammals. More recently Macdonald

Table 4 Some handbooks and faunal treatments of European mammals

Geographic region	Source
Austria	Rebel (1933), Spitzenberger (2001), and Stüber et al. (2014)
Belgium	Frechkop (1958)
Benelux	Saint Girons (1973)
British Islands	Millais (1904), Barrett-Hamilton (1910–1921), Thorburn (1920), Harrison Matthews (1952), Corbet and Southern (1964), Lawrence and Brown (1967), Arnold (1993), Harris et al. (1995), Macdonald and Tattersall (2001), and Harris and Yalden (2008)
Bulgaria	Markov (1957) and Peshev et al. (2004)
Czech Republic + Slovakia	Pelikán et al. (1979), Anděra (2000), Anděra and Horáček (1982), Anděra and Hanzal (1995, 1996), Anděra and Beneš (2001), and Anděra and Červený (2009)
Denmark	Baagoe and Jensen (2007)
Estland	Grevé (1909)
Estonia	Masing (1999)
Finland	Siivonen (1968, 1972)
France	Didier and Rode (1935), Rode and Didier (1946), Saint Girons (1973), Brosset (1974), and Fayard (1984)
Germany	Blasius (1857), Schäff (1911), Krumbiegel (1931), Mohr (1950), Haltenorth (1955), Angermann and Hackethal (1974), Herrmann (1991; Saarland), Borkenhagen (2011; Schleswig-Holstein), Hauer et al. (2009; Saxony), Görner (2009; Thuringia), Braun and Dieterlen (2003, 2005, Baden-Württemberg), and Grimmberger (2013)
Italy	Toschi (1965), Toschi and Lanza (1959), Spagnesi et al. (2000), and Spagnesi and De Marinis (2002)
Liechtenstein	von Lehmann (1963) and Broggi et al. (2011)
Lithuania	Balciauskas et al. (1999)
Macedonia	Kryštufek and Petkovski (2003)
Netherlands	Wijngaarden et al. (1971) and Broekhuizen et al. (1992)
Poland	Kowalski (1964), Pucek (1983), and Pucek and Raczynski (1983)
Romania	Murariu (2000), Popescu and Murariu (2001), and Murariu et al. (2016)
Slovakia	Krištofík and Danko (2012)
Slovenia	Kryštufek (1991)
Spain (Balearic Islands)	Alcover (1988)
Spain (mainland)	Graells (1897), Cabrera (1914), Vericad (1972), Gosàlbez i Noguera (1987), Rodríguez (1993), and Palomo and Gisbert (2002)
Switzerland	Baumann (1949), Rahm (1976), Hausser and Bourquin (1988), Hausser (1995), Marchesi et al. (2008), and Müller et al. (2010)
Yugoslavia (former)	Petrov (1992)

For books treating Russia and Far Eastern European countries, see text

(1995), Dietz et al. (2007), Temple and Terry (2007), Aulagnier et al. (2007), Grimmberger and Rudloff (2009), Twisk et al. (2010), and Dietz and Kiefer (2014; only bats) reviewed the status and distribution of European mammals. Mitchell-Jones et al. (1999) presented an atlas of European mammals as the result of an international cooperation. Temple and Cuttelod (2008) published a review of mammals of the Mediterranean area. Books treating mammals of mainly western Europe at a more local level are listed in Table 4. Wilson and Reeder (2005) and Wilson et al. (2009–2018) are modern checklists or handbooks which also include the European species.

The mammalogists of eastern Europe, which lived for the major part of the twentieth century within the borders of the Soviet Union (particularly Russia, Belarus, Ukraine, and Moldavia), published uncountable articles and books, often as part of wider geographical treatments. This includes a series of taxonomic compilations of various mammalian groups of Europe and

Palaearctic Asia, for example, of insectivores by Gureev (1979) and Zaytsev et al. (2014), a large number of works covering various rodents and lagomorphs which were summarized by Gromov and Erbajeva (1995), carnivorans and ungulates by Heptner and co-workers (e.g., Heptner and Sludskii 1992) which appeared in four volumes and so forth. There were several attempts to compile the entire mammal fauna of the Soviet Union under a single title, for example, by Bobrinskii et al. (1944 and reprinted editions), Gromov et al. (1963, in two volumes), Flint et al. (1965), and Pavlinov et al. (2002). The majority of publications during the Soviet period were in Russian and were therefore not accessible to the majority of mammologist working outside the Soviet Union. Because of their outstanding importance, some were translated into English and the best known in the West were seven volumes of the "Mammals of Eastern Europe and Northern Asia" (later volumes appeared under the title "Mammals of the USSR and Adjacent Countries") by Ognev published in the Soviet Union during 1928–1950 and released in English from 1962–1966 (Ognev 1962–1966; for references see Bakloushinskaya et al. 2012). Mammals are also covered in a large number of regional works. In the European part of Russia, Stroganov (1949) and Ivanter (2009) wrote about the mammals of Karelia, Estaf'ev (1994, 1998) of the extreme north-eastern European Russia, Vechkanov et al. (2004) about Mordovia, Kruskop (2002) about the Moscow area, Popov (1960) and Schlyakhtin et al. (2009) on the Volga region, Bol'shakov et al. (2000) of the Ural Mts, and so forth. Similarly, Serzhanin (1955 and reprinted editions) and Kozlo (2003) compiled knowledge on the mammals of Belarus, Dulitskiy (2001) about the Crimea, and Migulin (1938), Tatarinov (1956), Mezhzherin and Lashkova (2013), and Abelentsev with co-workers (in three volumes) about the Ukraine (zagorodniuk 2017).

The Discovery of Species

The cumulative number of acknowledged mammalian species in Europe has increased steadily ever since the 1758 Systema Naturae. The cumulative curve was the steepest before 1850, that is, during the period of most intensive naming of new species. Although the pace of new discoveries or descriptions slowed down after 1950, the curve is not yet asymptotic; hence, discoveries of new species are still likely. The curve reflects the development as perceived from the current state of knowledge (Fig. 1). As we already saw, the

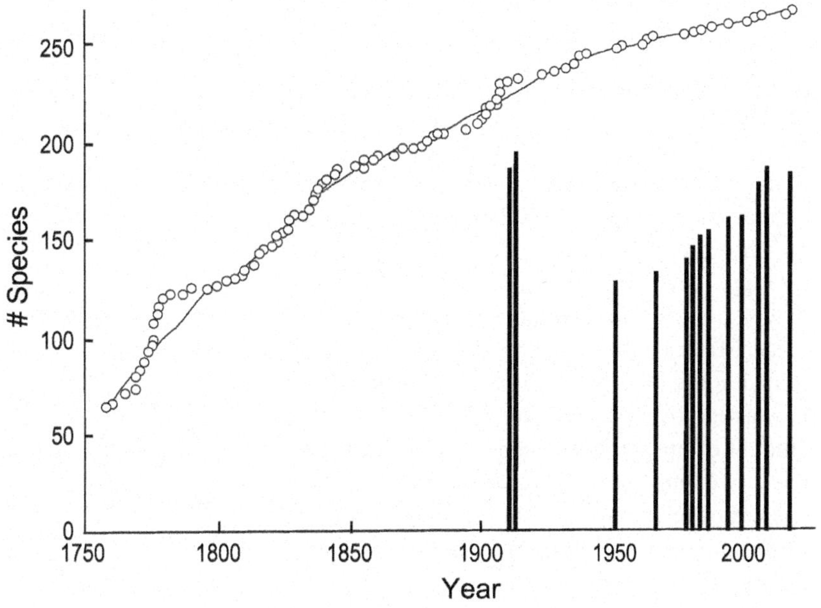

Fig. 1 Cumulative number of species of European mammals (dots) as recognized currently since the 1758 publication by Linnaeus. Note that the best-fit curve still does not reach the asymptote. Bars show variation in number of recognized species in Western Europe (i.e., Europe without Russia/Soviet Union) since the revisions of Trouessart (1910) and Miller (1912). Some species have been omitted to make different sources comparable

actual progress was much less directional, and therefore more erratic. In Western Europe, the number of recognized species was the highest in the early twentieth century but reached the lowest point in mid-century in a taxonomic revision by Ellerman and Morrison-Scott (1951). This work, which was at the time celebrated as "a magnificent synthesis ... [which] has sweepingly arranged the mammals of Eurasia ..." (Mayr 1963), is now denounced as representing a period of "taxonomic inertia" by some which underestimated the species richness, retarded the taxonomic progress in Europe, and in consequence affected biodiversity conservation policies (e.g., Gippoliti and Groves 2018). For a short treatment of different taxonomic philosophies, see Hackländer and Zachos (this volume). In the decades to follow Ellerman and Morrison-Scott (1951), the mammalogists were steadily adding "new" species (Table 1). In many cases they were arguing that taxa which already had valid names but were suppressed as junior synonyms do in fact represent full species. The new persuasive evidence frequently came from cytological laboratories, and later on from molecular evidence. That said, even in the light of new high-resolution molecular data, a large part of the disagreement over species delimitation and species numbers is due to fundamental differences in taxonomic philosophy, that is, which species concept one should follow.

Outlook

Today mammalogy is a complex science which is connected to a large set of other disciplines, such as physiology, cytology, ecology, population biology, behavior, conservation, morphology, paleontology, evolutionary biology, and so forth. Many of these fields were not even mentioned in the text. The basic disciplines however are taxonomy and evolution. Although a large amount of progress has been made in the recent past, we are still discovering and sorting species of mammals, also in Europe, attempting to put order into the mammal diversity that we find in nature.

References

Alcover JA (1988) Els mamífers de les Balears. Editorial Moll, Palma de Mallorca

Amori G, Castiglia R (2018) Mammal endemism in Italy: a review. Biogeogr J Integr Biogeogr 33:19–31

Anděra M (2000) Atlas of the mammals of the Czech Republic. A provisional version. Insectivores (Insectivora). Národni muzeum, Praha

Anděra M, Beneš B (2001) Atlas of the mammals of the Czech Republic. A provisional version. IV. Rodents (Rodentia). – Part 1. Hamsters (Cricetidae), voles (Arvicolidae), dormice (Gliridae). Národni muzeum, Praha

Anděra M, Červený J (2009) Large mammals in the Czech Republic. Distribution, history and protection. 1. Even-toed ungulates (Artiodactyla). Národní muzeum, Praha

Anděra M, Hanzal V (1995) Atlas of the mammals of the Czech Republic. A provisional version. I. Even-toed ungulates (Artiodactyla), lagomorphs (Lagomorpha). Národni muzeum, Praha

Anděra M, Hanzal V (1996) Atlas of the mammals of the Czech Republic. A provisional version. II. Carnivores (Carnivora). Národni muzeum, Praha

Anděra M, Horáček I (1982) Poznàváme nase savce. Mladà fronta, Praha

Angermann R, Hackethal H (1974) Säugetiere – Mammalia. In: Stresemann E (ed) Exkursionsfauna für die Gebiete der DDR und der BRD. Wirbeltiere. Volk und Wissen VEB, Berlin

Arnold HR (1993) Atlas of mammals in Britain. HMNS, London

Aulagnier S, Haffner P, Mitchell-Jones AJ, Moutou F, Zima J (2007) Mammifères d'Europe, d'Afrique du Nord et du Moyen-Orient. Delachaux & Niestle, Paris

Baagoe HJ, Jensen TS (2007) Dansk pattedyr Atlas. Nordisk Forlag A/S, Kobenhavn

Bakloushinskaya I, Bazykin GA, Kryukov AP, Kryukova EP (2012) Sergey Ivanovitch Ognev and the formation of theriology in Russia. Mammalia 76:375–382

Balciauskas L, Trakimas G, Juskaitis R, Ulevicius A, Balciauskiené L (1999) Atlas of Lithuanian mammals, amphibiens & reptiles, 2nd edn. Akstis Publications, Vilnius

Barrett-Hamilton GEH (1910–1921) A history of British mammals. Gurrney & Jackson, London

Barrett-Hamilton GEH (1913) A history of British mammals. Part XIV. Gurney and Jackson, London

Baumann F (1949) Die freilebenden Säugetiere der Schweiz. H Huber, Bern

Bjärvall A, Ullström S (1986) The mammals of Britain and Europe. Croom Helm, London/Sydney

Blasius JH (1857) Fauna der Wirbelthiere Deutschlands und der angrenzenden Länder von Mitteleuropa. I Vieweg & Sohn, Braunschweig

Bobrinskoy N, Kuznetzov B, Kuzyakin A (1944) Mammals of USSR. Sovietskaya Nauka, Moscow. (in Russian)

Bol'shakov VN, Berdyugin KI, Vasileva IA, Kuznetsova IA (2000) Mammals of Sverdlovsk Oblast. Ekaterinburg Press, Ekaterinburg. (in Russian)

Borkenhagen P (2011) Die Säugetiere Schleswig-Holsteins. Husum Druck- und Verlagsgesellschaft, Husum
Braun M, Dieterlen F (eds) (2003) Die Säugetiere Baden-Württembergs, vol 1. E Ulmer, Stuttgart
Braun M, Dieterlen F (eds) (2005) Die Säugetiere Baden-Württembergs, vol 2. E Ulmer, Stuttgart
Broekhuizen S, Hoekstra B, van Laar V, Smeenk C, Thissen JBM (1992) Atlas van de Nederlandse zoogdieren. Stichting Uitgeverij Koninklijke Nederlandse Natuurhistorische Vereniging, Arnheim
Broggi MF, Camenisch D, Fasel M, Güttinger R, Hoch S, Müller JP, Niederklopfer P, Staub R (2011) Die Säugetiere des Fürstentums Liechtenstein (Mammalia). Berichte der Botanisch-Zoologischen Gesellschaft Liechtenstein-Sargans-Werdenberg 36:5–200
Brosset A (1974) Mammifères sauvages de France et d'Europe de l'Ouest. Nathan
Burgin CJ, Coella JP, Kahn PL, Upham NS (2018) How many species of mammals are there? J Mammal 99:1–14
Cabrera A (1914) Fauna Ibérica. Mamíferos. Museo Nacional de Ciencias Naturales, Madrid
Corbet GB (1966) The terrestrial mammals of western Europe. G T Foulis & Co., London
Corbet GB (1978) The mammals of the Palaearctic Region: a taxonomic review. British Museum (Natural History) and Cornell University Press, London/Ithaca
Corbet GB (1984) The mammals of the Palaearctic Region: a taxonomic review. Supplement. British Museum (Natural History), London
Corbet GB, Ovenden D (1980) The mammals of Britain and Europe. Collins Sons & Co, London
Corbet GB, Southern HN (1964) The handbook of British mammals, 2nd edn. Blackwell Science, Oxford
Curry-Lindahl K (1975) Däggdjur i färg. Alla Europas arter. Almquist & Wiksell, Stockholm
Denys C, Callou C, Mellier B (2012) The contribution of Edouard-Louis Trouessart to mammalogy. Mammalia 76:355–364
Didier R, Rode P (1935) Les mammiféres de France. Société Nationalé d'Acclimatation de France, Paris
Dietz C, Helversen O, Nill D (2007) Handbuch der Fledermäuse Europas und Nordwestafrikas. Kosmos, Stuttgart
Dietz C, Kiefer A (2014) Die Fledermäuse Europas kennen, bestimmen schützen. Kosmos, Stuttgart
Dulitskiy AI (2001) Mammals of Crimea. Crimean Scientific and Pedagogical State Publisher, Simferopol. (in Russian)
Dunnum JL, Cook JA (2012) Gerrit Smith Miller: his influence on the enduring legacy of natural history collections. Mammalia 76:365–373
Ellerman JR, Morrison-Scott TCS (1951) Checklist of Palaearctic and Indian mammals 1758 to 1946. Trustees of the British Museum (Natural History), London
Ellerman J R, Morrison-Scott T C S. 1966. Checklist of Palaearctic and Indian mammals 1758 to 1946. 2nd edition. London: Trustees of the British Museum (Natural History)
Estaf'ev AA (ed) (1994) Fauna of the European north-eastern Russia. Mammals 1. Insectivores, bats, lagomorphs, rodents. Nauka, Sankt-Peterburg. (in Russian)
Estaf'ev AA (ed) (1998) Fauna of the European north-eastern Russia. Mammals 2. Cetaceans, carnivores, pinnipeds, artiodactyls. Nauka, Sankt-Peterburg. (in Russian)
Fayard A (1984) Atlas des mammifères sauvages de France. Société Francaise pour l'Étude et la Protection des Mammifères, Paris
Flint VE, Chugunov Yu D, Smirin VM (1965) Mammals of the USSR. Mysl, Moscow. (in Russian)
Frechkop S (1958) Faune de Belgique. Mammifères. Institut royal des Sciences naturelles de Belgique, Bruxelles
Gaffrey G (1961) Merkmale der wildlebenden Säugetiere Mitteleuropas. Akademische Verlagsgesellschaft, Leipzig
Genovesi P, Bacher S, Kobelt M, Pascal M, Scalera R (2009) Alien mammals of Europe. In: Nentwig W (ed) DAISIE, Handbook of Alien species in Europe. Springer, Dordrecht, pp 119–128
Genoways HH, Schlitter DA (1981) Collections of recent mammals of the World, exclusive of Canada and the United States. Ann Carnegie Museum 50:47–80
Gessner C, Forer C (1563) Thierbuch. Das ist ein kurtze beschreybung aller vierfüssigen Thieren/so auff der erde und in wassern wohnend/sampt irer waren konterfactur: alles zu nutz von gutem allen liebhabern der künsten/ Artzeten/Maleren/Bildschnitzern/Weydleuten und Köchen/gestelt. Christoffel Froschower, Zürich
Gippoliti S, Groves CP (2018) Overlooked mammal diversity and conservation priorities in Italy: impacts of taxonomic neglect on a Biodiversity Hotspot in Europe. Zootaxa 4434(3):511
Görner M (ed) (2009) Atlas der Säugetiere Thüringens. Biologie, Lebensräume, Verbreitung, Gefährdung, Schutz. Arbeitsgruppe Artenschutz Thüringen, Jena
Görner M, Hackethal H (1988) Säugetiere Europas. Neumann Verlag, Leipzig/Radebeul
Gosàlbez i Noguera J (1988) Insectívors; rosegadors de Catalunya. Ketres Editora, Barcelona
Graells M de la Paz (1897) Fauna Mastodológica Ibérica. Mem Real Acad Cienc Exact Fís Natur Madrid 17:1–806, 22 pls, 2 maps
Grevé K (1909) Säugetiere Kur- Liv- Estlands. W Mellin, Riga
Grimmberger E (2013) Die Säugetiere Deutschlands: Beobachten und Bestimmen. Quelle & Meyer, Wiebelsheim
Grimmberger E, Rudloff K, Kern C (2009) Atlas der Säugetiere Europas, Nordafrikas und Vorderasiens. Natur und Tier-Verlag, Münster
Gromov IM, Erbajeva MA (1995) The mammals of Russia and adjacent territories. Lagomorphs and Rodents. Russian Academy of Sciences, Zoological Institute, St. Petersburg. (in Russian)
Gromov I M, Gureev A A, Novikov G A, Sokolov I I, Strelkov P P, Chapskij K K. (1963) Mammals in the Fauna of the USSR. Parts I and II. Moscow: Academy of Sciences of the USSR. (in Russian)

Gureev AA (1979) Fauna of the USSR. Mammals, IV/2, insectivores (Mammalia, Insectivora). Hedgehogs, moles and shrews (Erinaceidae, Talpidae, Soricidae). Nauka, Leningrad. (in Russian)

Hainard R (1948/49) Les mammifères sauvage d'Europe. I, II. Delachaux & Niestlé, Neuchâtel/Paris

Haltenorth T (1955) Mammalia. In: Döderlen L (ed) Bestimmungsbuch für deutsche Land- und Süßwassertiere. 2. erw. Auflage. Band 4, Wirbeltiere. R. Oldenburg, München

Harris S, Yalden DW (2008) Mammals of the British Isles: handbook, 4th edn. The Mammal Society, Southampton

Harris S, Morris P, Wray S, Yalden D (1995) A review of British mammals: population estimates and conservation status of British mammals other than cetaceans. Joint Nature Conservation Committee, Peterborough

Harrison Matthews L (1952) British mammals. Collins, London

Hauer S, Ansorge U, Zöphel U (2009) Atlas der Säugetiere Sachsens. Sächsisches Landesamt für Umwelt, Landwirtschaft und Geologie, Dresden

Hausser J (1995) Säugetiere der Schweiz. Verbreitung, Biologie, Ökologie. Birkhäuser, Basel/Boston/Berlin

Hausser J, Bourquin J-D (1988) Atlas der Säugetiere der Schweiz. Die Verbreitung von zwölf Säugetierarten in der Schweiz. Schweizerische Gesellschaft für Wildforschung, Lausanne

Heptner VG, Sludskii AA (1992) Mammals of the Soviet Union. Volume II, Part 2. Carnivora (Hyaenas and Cats). Smithsonian Institution Libraries and National Science Foundation, Springfield

Herrmann M (1991) Säugetiere im Saarland. Verbreitung, Gefährdung, Schutz. Schriftenreihe des Naturschutzbundes Saarland e.V, St. Wendel

Hoffmeister DF, Sterling KB (1994) Origin. In: Birney EC, Choate JR (eds) Seventy-five years of mammalogy (1919–1994). American Society of Mammalogy, pp 1–21

Hutterer R (2001) Berlin und die Deutsche Gesellschaft für Säugetierkunde. Bongo (Berlin) 31:97–120

Ivanter EV (2008) Mammals of Karelia. PetrGU, Petrozavodsk. (in Russian)

Jaarola M, Searle JB (2002) Phylogeography of field voles (*Microtus agrestis*) in Eurasia inferred from mitochondrial DNA sequences. Mol Ecol 11(12):2613–2621

Johnston H (1903) British mammals: an attempt to describe and illustrate the Mammalian Fauna of the British Islands from the commencement of the Pleistocene Period to the present day. Woburn Library, London

Keyserling AG, Blasius JH (1840) Die Wirbelthiere Europa's. Friedrich Viehweg & Sohn, Braunschweig

König C (1969) Wildlebende Säugetiere Europas. Chr. Belser, Stuttgart

Kowalski K (1964) Klucze do oznaczania kregowców Polski. Ssaki-Mammalia. Polska Akademia Nauk, Krakow

Kozlo PG (ed) (2003) Mammals. Belarus Encyclopedia (in Belarus)

Krištofík J, Danko Š (2012) Mammals of Slovakia: distribution, bionomy and protection. Veda, Bratislava. (in Slovak with English summary)

Krumbiegel I (1931) Mammalia. In: Schulze P (ed) Biologie der Tiere Deutschlands. Gebrueder Borntraeger, Berlin

Kryštufek B (1991) Mammals of Slovenia. Prirodoslovni muzej Slovenije, Ljubljana. (in Slovene with English summary)

Kryštufek B, Petkovski S (2003) Annotated checklist oft he mammals of the Republic of Macedonia. Bonner Zoologische Beiträge 51:229–254

Kryštufek B, Tesakov AS, Lebedev VS, Bannikova AA, Abramson GI, Shenbrot G (2019) Back to the future: the proper name for red-backed voles is *Clethrionomys* Tilesius and not *Myodes* Pallas. Mammalia, online 09 Aug 2019

Lange R, Twisk P, van Winden A, van Diepenbeek A (1994) Zoogdieren von West-Europa. KNNV-uitgeverij, Utrecht

Lawrence MJ, Brown RW (1967) Mammals of Britain. Their tracks, trails and signs. Blanford Press, London

Lidicker WZ Jr (2011) The International Federation of Mammalogists: what is it and what does it do? Therya 2(1):5–6

Linnaeus C (1758) Systema naturae per regna tria naturae, secundum classes, ordines, genera, species, cum characteribus, differentiis, synonymis, locis. Laurentii Salvii, Holmiae. Vol. Tomus I, Editio decima, reformata, pp i–ii, 1–824

Lydekker R (1896) A hand-book to the British Mammalia. Edward Llyod, London

Macdonald D (1995) Collins European mammals. Evolution and behavior. Harper Collins, London

Macdonald D, Barrett P (1993) Mammals of Britain & Europe. Collins Field Guide. Harper Collins, London

Macdonald D, Tattersall F (2001) Britain's mammals: The challenge for conservation. People's Trust for Endangered Species, London

Marchesi P, Blant M, Capt S (2008) Fauna Helvetica 22. Säugetiere der Schweiz. Bestimmungsschlüssel. CSCF & SGW, Neuchâtel

Markov G (1957) (Insect eating mammals of Bulgaria). Fauna Bulgarica 3. Bulgarian Academy of Science, Zoological Institute, Sofia

Masing M (1999) Taxonomy and status of wild mammals in Estonia. Sicista, Tartu

Mayr E (1963) Animal species and evolution. Harvard University Press, Cambridge, MA

Mezhzherin SV, Lashkova OI (2013) Mammals of Ukraine. Naukova Dumka, Kiev. (in Ukrainian)

Migulin OO (1938) Mammals of the Ukrainian SSR (Data on the fauna). Ukrainian SSR Academy of Sciences Press, Kiev. (in Ukrainian)

Millais JG (1904) The mammals of Great Britain and Ireland. Longmans Green & Co., London

Miller GS (1912) Catalogue of the mammals of western Europe (Europe exclusive of Russia) in the collection

of the British Museum. British Museum (Natural History), London
Mitchell-Jones AJ, Amori G, Bogdanowicz W, Kryštufek B, Reijnders PJH, Spitzenberger F, Stubbe M, Thissen JBM, Vohralík V, Zima J (eds) (1999) The atlas of European mammals. Poyser Natural History, London
Mohr E (1950) Die freilebenden Nagetiere Deutschlands und der Nachbarländer. G. Fischer, Jena
Müller JP, Jenny H, Lutz M, Mühletaler E, Briner T (2010) Die Säugetiere Graubündens, eine Übersicht. Desertina, Chur
Murariu D (2000) Fauna Romaniei. Mammalia. Vol. XVI Fasc. 1 Insectivora. Ed. Acad. Romane, Bucuresti
Murariu D, Chişamera G, Măntoiu DŞ, Pocora I (2016) Romanian Fauna. Mammalia, 16(3) Chiroptera. Romanian Academy, Bucharest. (in English)
Niethammer J, Krapp F (1978–1992) Handbuch der Säugetiere Europas, vol 1–6. Akademische Verlagsgesellschaft & AULA-Verlag, Wiesbaden
Ognev SI (1962–1966) Mammals of U. S. S. R. and adjacent countries, vols 1–7. Israel Program for Scientific, Jerusalem
Osgood WH (1944) Bibliographical memoire of Clinton Hart Merriam 1855–1942. Bibliogr Memoir Nat Acad Sci U S A 24:1–57
Palomo LJ, Gisbert J (eds) (2002) Atlas de los mamíferos terrestres de Espana. Dirección General de Conservación de la Naturaleza – SECEM-SECEMU, Madrid
Pavlinov IY, Kruskop SV, Varshavskij AA, Borisenko AV (2002) Terrestrial mammals of Russia. KMK, Moscow. (in Russian)
Pelikán J, Gaisler J, Rödl P (1979) Nasi savci. Albatros, Nakladatel
Peshev TH, Peshev DT, Popov VV (2004) The fauna of Bulgaria. 27. Mammalia. Edition Academia "Marin Drinov", Sofia
Petrov BM (1992) Mammals of Yugoslavia. Insectivores and rodents. Nat Hist Museum Belgrade (Supplementa Special Issues) 37:1–186
Popescu A, Murariu D (2001) Romania's Fauna, Mammalia16 (2), Rodentia. Bucureşti, Editura Academiei Române. (in Romanian)
Popov VA (1960) Mammals of the Volga-Kama region. Insectivores, bats, rodents. Academy of Sciences of the USSR, Kazan branch, Kazan. (in Russian)
Pucek Z (1983) Keys to vertebrates of Poland. Mammals. PWN-Polish Scientific Publishers, Warszawa
Puzek Z, Raczynski J (1983) Atlas rozmieszczenia ssaków W Plsce mapy. Polska Akademia Nauk, Zaklad Badania Ssakow. PWN-Polish Scientific Publishers, Warszawa
Rahm U (1976) Die Säugetiere der Schweiz. Veröffentlichungen aus dem Naturhistorischen Museum Basel 9:1–87
Rebel H (1933) Die freilebenden Säugetiere Österreichs als Prodromus einer heimischen Mammalienfauna. Österreichischer Bundesverlag für Unterricht, Wissenschaft und Kunst, Wien/Leipzig
Rode P, Didier R (1946) Atlas des mammifères de France. N. Boubée & Cie, Paris
Rodríguez JL (1993) Guía de campo de los mamíferos terrestres de Espana. Ediciones Omega, Barcelona
Saint Girons M-C (1973) Les mammifères de France et de Benelux (fauna marine exceptée). Doin, Paris
Schäff E (1911) Die wildlebenden Säugetiere Deutschlands. J. Neumann, Neudamm
Schiebinger L (1993) Nature's body: gender in the making of modern science. Beacon-Press, New Brunswick
Schilling D, Singer D, Diller H (1983) BLV Bestimmungsbuch Säugetiere. 181 Arten Europas. BLV, München
Schmiedeknecht O (1906) Die Wirbeltiere Europa's mit Berücksichtigung der Faunen von Vorderasien und Nordafrika. Fischer, Jena
Serzhanin IN (1955) Mammals of the Belarus Soviet Socialist Republic. Academy of Sciences of the Belarus Soviet Socialist Republic, Minsk. (in Russian)
Shamel HH, Miller AG, Schwarz E, Clark AH, Bartsch P, Jackson HHT, Wright AH, Howell AB, Rehn JAG, Miller GS Jr, Johnson DH (1954) Gerrit Smith Miller, Jr. J Mammal 35:317–344
Shlyakhtin GV, Il'in VY, Oparin ML, Beljachenko AV, Bystrakova NV, Ermakov OA, Zav'jalov EV, Zakharov KS, Kaybeleva EI, Koshkin VA, Kurmaeva NM, Luk'yanov SB, Mosolova EY, Oparina OS, Semikhatova SN, Smirnov DG, Sonin KA, Tabachishin VG, Titov SV, Filip'echev AO, Khuchraev SO, Yakushev NN (2009) Mammals of the Northern Lower-Volga region. Vol. I. Theriofauna composition. Saratov University Press, Saratov. (in Russian, English abstract)
Siivonen L (1968) Nordeuropas däggdjur. P. A. Norstedt & Söners, Stockholm
Siivonen L (1972) Suomen nisäkkäät 1. Helsingissä kustannus osakeyhtiö, Otavo
Spagnesi M, de Marinis AM (eds) (2002) Mammiferi d'Italia. Quaderni di Conservacione della Natura 14:1–309
Spagnesi M, Toso S, de Marinis A M (eds) (2000) Italian mammals. Ministerio dell'Ambiente: Servicio Conservazione della Natura and Istituto Nazionale per la Fauna Selvativa A Ghigi
Spitzenberger F (2001) Die Säugetierfauna Österreichs. Grüne Reihe des Bundesministeriums für Land- und Forstwirtschaft, Umwelt und Wasserwirtschaft, Graz
Stroganov SU (1949) Key to the Mammals of Karelia. Karelian-Finnish State University, Petrozavodsk. (in Russian)
Stüber E, Lindner R, Jerabek M (2014) Die Säugetiere Salzburgs. Haus der Natur, Salzburg
Tatarinov KA (1956) Mammals of western parts of Ukraine. Academy of Sciences of the Ukrainian Soviet Republic, Kiev. (in Ukrainian)
Temple HJ, Cuttelod A (2008) The status and distribution of Mediterranean mammals. IUCN, Gland/Cambridge

Temple HJ, Terry A (2007) The status and distribution of European mammals. Office for Official Publications of the European Communities, Luxembourg

Thorburn A (1920) British mammals, vols I, II, London, Longmans, Green & Co

Toschi A (1965) Fauna d'Italia. Mammalia. Lagomorpha. Rodentia. Carnivora – Artiodactyla. Cetacea. Edizioni Calderini, Bologna

Toschi A, Lanza B (1959) Fauna d'Italia. Mammalia. Generalità, Insectivora. Chiroptera. Edizioni Calderini, Bologna

Trouessart E-L (1910) Faune des mammifères d'Europe. Friedländer & Sohn, Berlin

Twisk P, van Diepenbeeck A, Bekker JP (2010) Veldgids Europese zoogdieren. Zeist, KNNV Uitgeverij

van den Brink FH (1955) Zoogdierengids van Europa ten westen van 30° oosterlengte. Elsevier, Amsterdam/ Bruxelles

van Wijngaarden A, van Laar V, Trommel MDM (1971) De verspreiding van de Nederlandse zoogdieren (Mammalia: Insectivora, Chiroptera, Rodentia, Lagomorpha, Carnivora, Pinnipedia, Artiodactyla). Lutra 13:1–42 + 64 maps

Vechkanov VS, Kuznetsov VA, Al'ba LD, Ruchin AB (2004) Mammals of Mordovia. University of Mordovia, Saransk. (in Russian)

Vericad JR (1972) Estudio faunístico y biológico de los mamíferos del Pirineo. Publicaciones de Centro Pirenaico de Biologia experimental 4:7–232 + 56 photos

von Lehmann E (1963) Die Säugetiere des Fürstentums Liechtenstein. Jahrbuch des Historischen Vereins im Fürstentum Liechtenstein 62:159–362

Wilson DE, Reeder DA (eds) (2005) Mammal species of the World. A taxonomic and geographic reference, 3rd edn. The Johns Hopkins University Press, Baltimore

Wilson DE, Lacher TE, Mittermeier RA (eds) (2009-2019) Handbook of the mammals of the world, vols 1–8. Lynx Edicions, Barcelona

Zagorodniuk I (2017) Vaysl Abelentsev (1913–1980): Life in the world of science. Proceedings of the theriological school 15:167–174 (in Ukrainian)

Zaytsev MV, Voyta LL, Sheftel BI (2014) Mammals in the fauna of Russia and adjacent regions. Insectivores. Russian Academy of Sciences. Institute of Zoology, Sankt-Petersburg. (in Russian)

Mammalian Phylogenetics: A Short Overview of Recent Advances

3

Frank E. Zachos

Contents

Abstract

This chapter summarizes our present understanding of the phylogenetic relationships of Mammalia, particularly those within Marsupialia and Placentalia. The last 20 or so years have seen the burgeoning of molecular phylogenetics and the transition from phylogenetics to phylogenomics, with new and deeper insights into mammalian relationships. While most of the taxa traditionally classified as "orders" have stood the test of time, the "interordinal" relationships have benefited immensely from the new methodology. This is most obvious for placental mammals where four high-ranking taxa have emerged beyond reasonable doubt: Afrotheria, Xenarthra, Euarchontoglires, and Laurasiatheria. The root of the placental tree, however, is still elusive, with a number of competing hypotheses still being discussed. The most likely topology seems to be a most basal split between Atlantogenata (= Afrotheria + Xenarthra) and Boreoeutheria (= Euarchontoglires + Laurasiatheria). While Boreoeutheria is well supported, this is much less the case for Atlantogenata. The position of Scandentia (tree shrews) and Chiroptera (bats) within Euarchontoglires and Laurasiatheria, respectively, is also still uncertain. Another heatedly debated issue is the age of origination and diversification of the placental mammals, particularly with respect to the K-Pg boundary ca. 66 mya. Molecular datings and the fossil record are still at odds with one another, but a reconciliation seems at least feasible.

F. E. Zachos (✉)
Mammal Collection, Natural History Museum Vienna, Vienna, Austria

Department of Evolutionary Biology, University of Vienna, Vienna, Austria

Department of Genetics, University of the Free State, Bloemfontein, South Africa
e-mail: frank.zachos@nhm-wien.ac.at

K. Hackländer, F. E. Zachos (eds.), *Mammals of Europe - Past, Present, and Future*, Handbook of the Mammals of Europe, https://doi.org/10.1007/978-3-030-00281-7_6

Keywords

K-Pg Boundary Marsupialia · Molecular Phylogenetics · Monotremata · Placentalia

Introduction

The present handbook's predecessor, the German *Handbuch der Säugetiere Europas*, was published between the late 1970s and early 2000s, i.e., largely before the period when, over the last 15–20 years, our views on the phylogenetic relationships among extant mammal groups have been shaken up by molecular studies. This is particularly true for the placental Tree of Life that is most relevant to Europe. Although the handbook is not primarily about phylogenetic relationships, we thought it appropriate to include a short chapter summarizing the revolutionary insights of the last two decades to put the species accounts into a wider evolutionary perspective. While the *Handbook of the Mammals of Europe* – being a zoological, not a paleontological reference work – only covers the extant mammalian fauna, it is also true that zoologists oftentimes tend to have a one-dimensional view of biota that is biased toward the present. There is nothing wrong with studying extant mammals, of course, but it should be kept in mind that in the past, mammalian biota looked very different, and those readers who wonder what the phylogeny of afrotherians or marsupials is to do with European mammals should remember that many higher taxa[1] that today only occur outside our continent used to be an essential part of its mammal fauna in earlier Cenozoic times and sometimes as recently as the last ice age. Rhinoceroses, wild horses, hippopotamuses and big cats, but also afrotherians (most famously mammoths) were part of the European Pleistocene mammal fauna. In the Miocene, Europe harbored great apes (e.g., *Dryopithecus*, Begun 1992, *Pierolapithecus*, Moyà-Solà et al. 2004, and *Danuvius guggenmosi*, Böhme et al. 2019) and the famous Eocene fossils from the Messel Pit in Germany include, among other "exotics," pangolins (*Eomanis*, e.g., Gaudin et al. 2009) and even marsupials (*Peradectes*, Wilson et al. 2016; *Peradectes* is sometimes considered to be a stem metatherian rather than within Marsupialia). That Messel's *Eurotamandua* is a European xenarthran (it was originally thought to be an anteater, Storch 1981) is now considered refuted; instead, it seems to belong within Pholidota (Gaudin et al. 2009) or closely related to the Palaeanodonta (Rose 1999). A comprehensive up-to-date review about mammalian phylogenetics including fossil taxa (with a focus on placentals) is given by Asher (2018). For recent reviews on mammalian roots among Paleozoic synapsids and on mammals in the Mesozoic, see Angielczyk and Kammerer (2018) and Martin (2018).

This chapter addresses the extant part of mammalian diversity (Fig. 1). There have been numerous studies recently, and I will therefore only give a concise summary of the present consensus and of open or contentious questions pertaining to the interrelationships and evolutionary history of higher taxa within Mammalia. While it necessarily contains an element of arbitrariness which clade along the synapsid lineage is assigned the name Mammalia (a common solution is based on the presence of the squamoso-dental, or secondary, jaw joint), this issue need not bother us here, since we are dealing with the mammalian crown group only – the least inclusive clade containing all extant mammals, i.e., the most recent common ancestor of monotremes, marsupials, and placentals, and all its descendants (living and fossil). I am aware of the different views as to how inclusive the clade Mammalia should be when including fossil taxa, but in a zoological context of extant taxa only, Mammalia is the same as crown-group Mammalia and hence unambiguous. I will therefore simply refer to the crown group as Mammalia in this chapter.

The two highest-ranking sister taxa within Mammalia are Monotremata and Theria, and the latter in turn comprises Metatheria and Eutheria. The only alternative that was seriously considered is a sister group relationship between Monotremata and Metatheria/Marsupialia (Marsupionta hypothesis, Gregory 1947, contra Gregory 1910; Janke et al. 2002). This is today considered refuted, also by those authors who earlier favored it, see Kullberg et al. 2008. Until recently, the monotremes along with fossil taxa along its stem

Fig. 1 (continued)

Fig. 1 (continued)

were called Prototheria. It should be made clear right from the start that these names are somewhat ill-chosen as the Greek prefixes proto- ("first"), meta- ("after, behind") and eu- ("good, true") imply a sequence from lowly, primitive Prototheria via slightly more advanced Metatheria to the most highly developed "true" mammals, the Eutheria. For logical reasons alone, no extant group can be more basal or lower than another. The interpretation of phylogenies as ladders of progress is a common misconception that has a long history ("scala naturae") and is often perpetuated but that is scientifically flawed and obsolete (e.g., Baum et al. 2005; Omland et al. 2008; Rigato and Minelli 2013; Zachos 2016), and mammals are no exception here (Weisbecker 2015). What is true, however, is that placental mammals are the dominant group in terms of species number and ecological space occupied. There are 5,000–6,000 extant placental mammal species compared to 300–400 marsupial and 5 monotreme species,[2] and only placental mammals have evolved powered flight (Chiroptera) and a complete secondarily aquatic lifestyle (Cetacea, Sirenia). Proto-, Meta-, and Eutheria are often used synonymously with Monotremata, Marsupialia, and Placentalia, respectively, but this is only correct when only extant groups are considered. Monotremes, marsupials, and placentals are the crown groups of Proto-, Meta- and Eutheria, but the latter also comprise their respective fossil stem taxa (Fig. 2).[3] As mentioned earlier, the term Prototheria is usually not used anymore, but for the present chapter, this group is largely irrelevant anyway.

Within Monotremata, the two highest-ranking sister taxa are the Ornithorhynchidae and the Tachyglossidae. The duck-billed platypus (*Ornithorhynchus anatinus*) is the only extant species of the former, while the latter comprises the short-beaked echidna (*Tachyglossus aculeatus*) and three species of long-beaked echidna (*Zaglossus* spp.). All *Zaglossus* species are today confined to New Guinea and are classified as Critically Endangered in the IUCN Red List.

Phylogenetic relationships within the Marsupialia and the Placentalia are much more complex and contentious, and it is here that, over the last 20 or so years, molecular phylogenetic studies have yielded novel insights and breakthroughs.

A number of molecular phylogenetic publications around the year 2000 in particular, dealing with higher taxa within Placentalia ("interordinal" relationships[4]), heralded the molecular age in mammalian phylogenetics (e.g., Springer et al. 1997; Stanhope et al. 1998; Madsen et al. 2001; Murphy et al. 2001a, b; Scally et al. 2001; the early findings are summarized in Springer et al. 2004, later reviews are given by Asher et al. 2009 and Foley et al. 2016). Two of the most widely discussed findings were (i) the inclusion of the whales (Cetacea) in the artiodactyls (in an extant sister group relationship with the Hippopotamidae)[5] and (ii) that what had been lumped into the wastebasket taxon "Insectivora" was actually a polyphyletic group of taxa

Fig. 1 Diversity of extant mammal groups. (**a**) Monotremata – a platypus (*Ornithorhynchus anatinus*) from the early nineteenth century collected by Ferdinand Bauer and now part of the Mammal Collection of the Natural History Museum Vienna (NMW 26513). (**b–f**) Marsupialia. (**b**) extinct thylacine, or Tasmanian tiger (*Thylacinus cynocephalus*), from the collection of the Natural History Museum Vienna (NMW ST 132); (**c**) monito del monte (*Dromiciops gliroides*); (**d**) koala (*Phascolarctos cinereus*); (**e**) quokka (*Setonix brachyurus*); (**f**) western gray kangaroo (*Macropus fuliginosus*). (**g-i**) Afrotheria (Placentalia). (**g**) rock hyrax (*Procavia capensis*); (**h**) African savanna elephant (*Loxodonta africana*); (**i**) lowland streaked tenrec (*Hemicentetes semispinosus*). (**j**) Xenarthra (Placentalia) – southern naked-tailed armadillo (*Cabassous unicinctus*). (**k–p**) Laurasiatheria (Placentalia). (**k**) leopard (*Panthera pardus*); (**l**) hippopotamus (*Hippopotamus amphibius*); (**m**) southern right whale (*Eubalaena australis*); (**n**) square-lipped or white rhinoceros (*Ceratotherium simum*); (**o**) epauletted fruit bat (*Epomophorus* sp.); (**p**) northern white-breasted hedgehog (*Erinaceus roumanicus*). (**q–s**) Euarchontoglires (Placentalia). (**q**) yellow baboon (*Papio cynocephalus*); (**r**) European ground squirrel (*Spermophilus citellus*); (**s**) common hamster (*Cricetus cricetus*). (Photo credits: All photos taken by the author except for (**a**) and (**b**) (courtesy Alice Schumacher, Natural History Museum Vienna), (**c, d, i**) (José Luis Bartheld, Sanjay ach and Frank Vassen, Wikimedia Commons, https://creativecommons.org/licenses/by/2.0/deed.en) and (**j**) (courtesy Arnaud Desbiez))

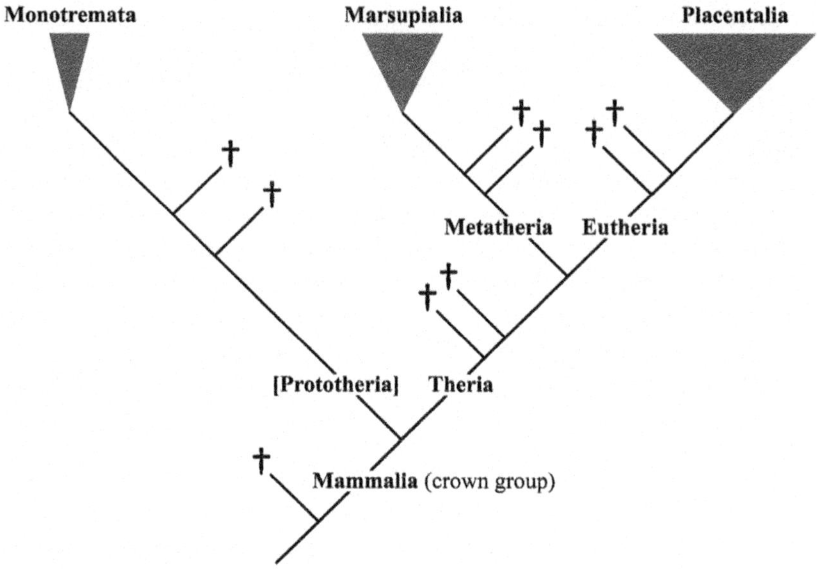

Fig. 2 Crown group Mammalia and less inclusive monophyletic groups. The sister group relationship within Mammalia of Metatheria/ Marsupialia and Eutheria/ Placentalia is well-corroborated. The term Prototheria is in parentheses as it is usually not used anymore (see text)

distributed throughout the newly emerging mammalian tree.

In the widest sense, "insectivorans" comprised the two groups named Menotyphla and Lipotyphla by Ernst Haeckel (based on the presence or absence, respectively, of a caecum). Menotyphla included tree shrews (Scandentia), elephant shrews (Macroscelidea), and colugos (Dermoptera), groups that are today regarded as belonging to very different major clades of mammals (see below). Lipotyphla comprised hedgehogs and relatives (Erinaceidae, Galericidae), shrews (Soricidae), moles (Talpidae), solenodons (Solenodontidae), and tenrecs and relatives (Tenrecidae, Potamogalidae, Chrysochloridae). The two milestone publications by Gregory (1910) and Simpson (1945) reduced the content of "Insectivora" but still carried the signature of this classification history. Gregory removed the Menotyphla from the insectivorans but kept all of the lipotyphlan taxa and coined the term Archonta for a group containing tree shrews, elephant shrews, colugos, bats, and primates. Simpson kept the elephant shrews within his "Insectivora" along with all the lipotyphlan taxa, classified colugos as a distinct "order," and included tree shrews in the Lemuriformes within the primates. The modern view is that tenrecs and relatives are part of an African placental clade called Afrotheria and that the classical insectivorans, now called Lipotyphla or Eulipotyphla, comprise hedgehogs and relatives, shrews, moles, solenodons, and, closely related to the latter, the recently extinct Caribbean *Nesophontes* (Brace et al. 2016).

While it is true that molecular phylogenetics has yielded new and even groundbreaking insights into the phylogeny of extant mammal taxa, particularly with regard to deep splits within Placentalia ("interordinal" relationships), it must be emphasized that the morphology-based taxa traditionally classified as "orders" are still considered valid clades (with few exceptions such as the "Insectivora"). The same holds for some "interordinal" groupings. For example, it has long been known that rodents and lagomorphs were closely related, and the combined taxon Glires is still valid. So is the close relationship among proboscideans, sirenians, and hyraxes, combinedly called Paenungulata (along with a number of fossil taxa). Comparing Novacek (1992), published 5 years before the molecular "revolution" in mammalian phylogenetics took off, with the most recent mammal trees (e.g., Fig. 4) gives a good idea of what has been overturned and what has stood the test of time. Morphological characters have also retrieved commonly accepted marsupial "orders" (e.g., Horovitz and Sánchez-Villagra 2003), and morphology-based relationships that were long considered false based on most molecular studies have more recently found support again (see the case of the marsupial mole *Notoryctes* below).

Marsupial Phylogeny

Although many more publications have dealt with the relationships of higher taxa within placental mammals, there has also been considerable interest in marsupial phylogeny and evolution. At present, there are 300–400 marsupial species, which is an order of magnitude less than the number for placentals. By and large, this discrepancy also holds in the fossil record (with notable exceptions such as the diverse marsupial fauna of South America before the Great American Interchange in the Pliocene). The fact that placental mammals exhibit so many more species (i.e., diversity) and also occupy larger eco- and morphospaces (i.e., disparity) has often been attributed to the peculiar reproductive biology of marsupials (early birth, long extra-uterine development, need for well-developed forelimbs at early ontogenetic stages which leaves less room for adaptive radiations). However, evidence for this explanation is not straightforward, and physiological and geographical reasons may be equally or more important. It seems that radiations on northern continents were much more successful in terms of species numbers than those on southern continents. This is not only true for the marsupial radiations vs. the placental ones, but also holds within the Placentalia where the two higher-level northern taxa Laurasiatheria and Euarchontoglires are almost 50 times more speciose than the two higher-level southern taxa Afrotheria and Xenarthra (see Sánchez-Villagra 2013 for a detailed discussion; for an explanation of the placental taxon names, see below).

A number of recent studies have addressed marsupial phylogeny (e.g., Asher et al. 2004; Nilsson et al. 2004; Meredith et al. 2008, 2011; Mitchell et al. 2014; May-Collado et al. 2015), and there are some clades that have been retrieved by most studies. May-Collado et al. (2015) give a list of and information on these "benchmark clades" along with references for them based on morphology and molecular data. They also provide a nice summary of the literature on marsupial phylogeny up to that time. Among the benchmark clades are high-level taxa such as Paucituberculata (shrew opossums), Didelphimorphia (New World opossums), Australidelphia (a large clade comprising all Australasian marsupials as well as the South American monito del monte *Dromiciops gliroides*), Dasyuromorphia (carnivorous Australian marsupials like the numbat, the Tasmanian devil and the extinct thylacine), and Diprotodontia (a speciose clade containing the koala, vombats, possums, cuscuses, kangaroos and wallabies, and others).

There is a consensus that the Australasian marsupials are nested within the American groups, but there are two alternative tree topologies for the deepest split within Marsupialia. Either the American opossums (Didelphimorphia) are sister to all other marsupials, or the deepest split is between the Paucituberculata and the rest:

(1) Marsupialia	(2) Marsupialia
Didelphimorphia	Paucituberculata
N.N.	N.N.
Paucituberculata	Didelphimorphia
Australidelphia	Australidelphia

Scenario (1) is supported by the studies by Asher et al. (2004), Nilsson et al. (2004), and Meredith et al. (2008), and it is also favoured by Eldridge et al. (2019), while more recent analyses favor Paucituberculata as the sister to all other marsupials (Meredith et al. 2011; Mitchell et al. 2014; May-Collado et al. 2015). The third possibility, a sister group relationship between Didelphimorphia + Paucituberculata (= Ameridelphia) on the one hand and Australidelphia on the other, has also not been refuted (Eldridge et al. 2019).

Within Australidelphia, the position of the South American *Dromiciops gliroides* (the monito del monte), the only extant species of Microbiotheria, is not fully resolved. While in most phylogenetic analyses it is the sister taxon of all other australidelphids (Eomarsupialia) (Meredith et al. 2008, 2011; Mitchell et al. 2014;

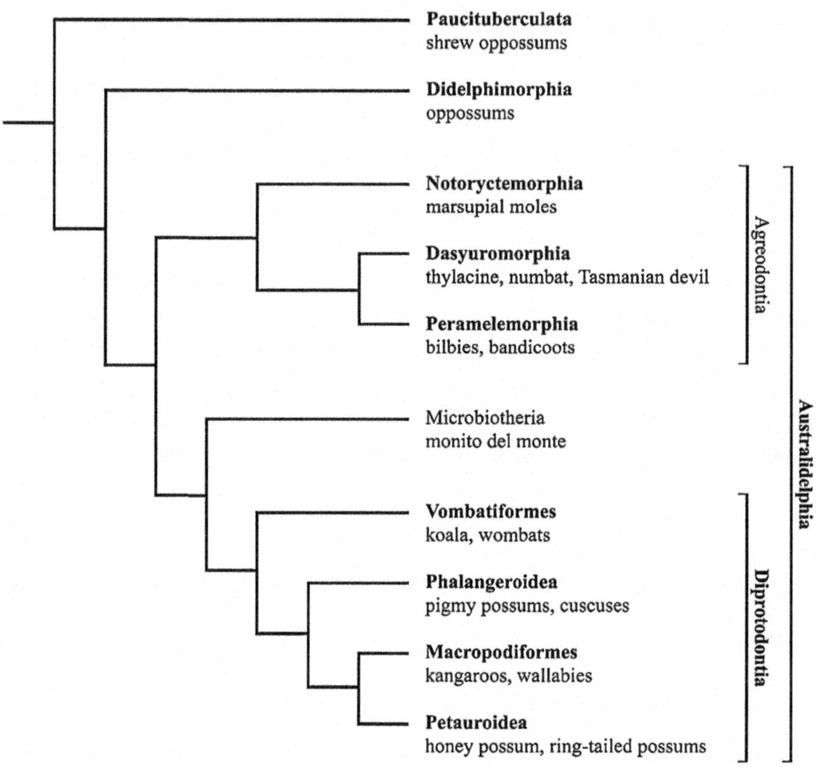

Fig. 3 Phylogenetic relationships of higher marsupial taxa based mainly on Fig. 1 in May-Collado et al. (2015). Morphological and some molecular studies favor the deepest split between Didelphimorphia (rather than Paucituberculata) and the remaining marsupials. The position of Microbiotheria also differs among studies. All named higher taxa in this phylogeny are listed as "benchmark clades" by May-Collado et al. and are in bold (Microbiotheria only comprises one extant species and does not qualify as a supraspecific clade). The colloquial names in parentheses are mostly only examples, not complete lists. The name Agreodontia for the clade comprising Notoryctemorphia, Dasyuromorphia, and Peramelemorphia was coined by Beck et al. (2014)

Duchêne et al. 2018), there is also evidence that the Microbiotheria are a lower-ranking taxon within the Australidelphia, potentially sister to the Diprotodontia (May-Collado et al. 2015). Asher et al. (2004), in a combined analysis of morphological and molecular data, also found the position of *Dromiciops* to be uncertain. The marsupial moles (*Notoryctes* sp., Notoryctemorphia) have usually been considered the sister taxon to Dasyuromorphia (Meredith et al. 2008) or Dasyuromorphia + Peramelemorphia (see Fig. 3). In a very recent phylogenomic study, however, Duchêne et al. (2018) found strong support for a sister group relationship of marsupial moles and Peramelemorphia, to the exclusion of Dasyuromorphia, in line with earlier morphological findings (Horovitz and Sánchez-Villagra 2003). Irrespective of the exact relationships among these three groups, the clade which they form has been considered supported well enough to deserve a name: Agreodontia (Beck et al. 2014; see Fig. 3).

Figure 3 shows the marsupial phylogeny based on the summary cladogram in May-Collado et al. (2015).

Placental Phylogeny

As mentioned above, most of the morphology-based higher taxa ("orders") of placentals are still valid. The two main areas of research with

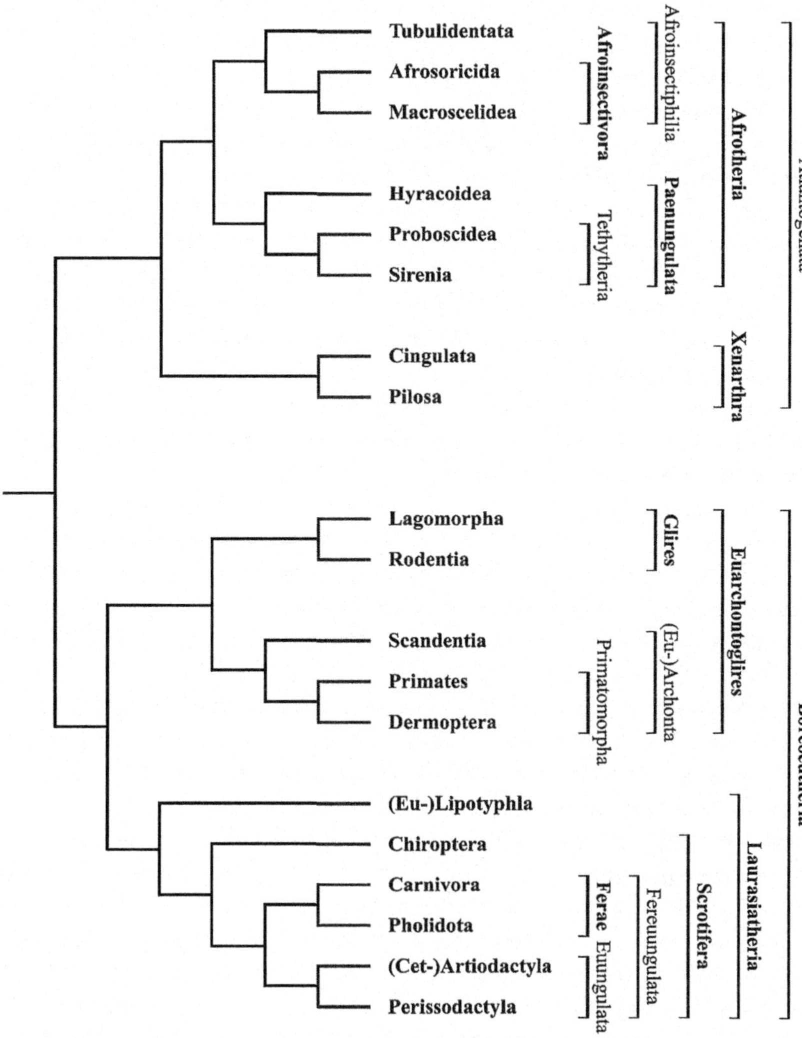

Fig. 4 Phylogenetic relationships of higher placental taxa down to the groups usually ranked as "orders" in Linnaean classifications. Combined after different studies cited in this chapter. See text for details and contentious splits. Taxa whose monophyly is well established are in bold ("benchmark clades")

respect to placental phylogeny are (i) the phylogenetic relationships themselves, i.e., the topology of the placental tree, and (ii) the temporal framework, i. e., the time of origin of various higher taxa and their evolutionary rates, particularly with respect to the K-Pg boundary ca. 66 mya (short fuse, long fuse, and explosive models, see below).

The major progress brought about by molecular phylogenetics over the last 20 or so years was the combination of higher-level placental taxa into four major clades that have repeatedly been corroborated and are accepted by most mammalian phylogeneticists. These four clades are:

1. Xenarthra (Cingulata, Pilosa)
2. Afrotheria (Proboscidea, Sirenia, Hyracoidea, Tubulidentata, Macroscelidea, Afrosoricida)
3. Euarchontoglires (Scandentia, Primates, Dermoptera, Rodentia, Lagomorpha)
4. Laurasiatheria ((Eu-)Lipotyphla, Chiroptera, Pholidota, Carnivora, (Cet-)Artiodactyla, Perissodactyla)[6]

Except for Xenarthra, these taxa are mainly based on molecular data, although for Afrotheria, an increase in the number of thoracolumbar vertebrae has been found to be a morphological apomorphy (Sánchez-Villagra et al. 2007). The interrelationships among these four groups and the root of the placental tree, i.e., which two sister taxa result from the first split within Placentalia, are less well-resolved. Most studies favor a sister group relationship of Euarchontoglires and Laurasiatheria (Boreoeutheria), but the position of Boreoeutheria relative to the remaining two clades is unresolved. There are still three alternatives being discussed according to which the placental root is between:

1. Xenarthra and Epitheria (= Afrotheria + Boreoeutheria)
2. Afrotheria and Exafroplacentalia (= Notolegia = Xenarthra + Boreoeutheria)
3. Boreoeutheria and Atlantogenata (= Xenarthra + Afrotheria)

Wildman et al. (2007) favored the Boreoeutheria/Atlantogenata hypothesis, linking placental evolution to plate tectonic events: the basal split between Boreoeutheria and Atlantogenata to the breakup of Pangaea into Gondwana (Atlantogenata) and Laurasia (Boreoeutheria) and the divergence of Afrotheria and Xenarthra to the separation of Africa and South America (the latter was also hypothesized by, for example, Murphy et al. 2001b, 2007). This was criticized, based on retroposon analysis and geological data, by Nishihara et al. (2009) who suggested a near-simultaneous separation of Laurasia, Africa, and South America and, concomitantly, also a near-simultaneous divergence of Boreoeutheria, Afrotheria, and Xenarthra. In any case, if continental vicariance through the opening of the South Atlantic was the driver of the divergence of afrotherians and xenarthrans, their most recent common ancestor must have lived prior to 100 mya (and Murphy et al. 2007 give an age of 103 mya for the Atlantogenata node). This, however, is not in accordance with more recent estimates based on larger datasets which have that node closer to 90 mya or even slightly younger (e.g., dos Reis et al. 2012; Tarver et al. 2016, and Wu et al. 2017). Asher (2018, p. 322f.) concludes from this that rather than continental vicariance, a "relatively narrow South Atlantic ocean between 83 and 96 Ma (Scotese 2001) [...] likely played an important role as a dispersal filter between populations that subsequently gave rise to Xenarthrans and Afrotherians." Tarver et al. (2016) emphasize that the dispersal of stem xenarthrans across the opening Atlantic ocean is less far-fetched than it might seem at first glance for two reasons. First, as they point out, this barrier was less significant than the one between Africa and Madagascar across which several post-Mesozoic dispersal events of placental mammals are known to have occurred (primates, rodents, tenrecs, carnivorans). Secondly, oceanic dispersal of Eocene rodents and primates across the by then much wider South Atlantic is uncontroversial (see, e.g., Bond et al. 2015).

With the exception of O'Leary et al. (2013a), the most comprehensive recent genomic studies have not found support for Xenarthra as sister to the remaining placentals (Epitheria hypothesis), but have nonetheless yielded contradictory results: while Romiguier et al. (2013) found evidence for Afrotheria as sister to all other placentals (Exafroplacentalia hypothesis), dos Reis et al. (2012), Morgan et al. (2013), Tarver et al. (2016), and Esselstyn et al. (2017) advanced the Boreoeutheria/Atlantogenata split (for a commentary see Teeling and Hedges 2013). Although these relationships are usually mainly analyzed based on molecular (and recently mostly genomic) data, a new cladistic morphological study on Paleocene placentals also retrieved the Boreoeutheria/Atlantogenata split (Halliday et al. 2017), adding additional weight to this scenario.

Figure 4 summarizes the potential placental phylogenetic relationships, assuming the Atlantogenata/Boreoeutheria hypothesis.

The monophyly of the terminal taxa in Fig. 4 ("orders") is well established and usually not contentious. The same holds for a number of higher taxa which are recovered by practically all comprehensive phylogenetic studies. Apart from Boreoeutheria, Xenarthra, Afrotheria, Euarchontoglires, and Laurasiatheria (see above), these taxa are Afroinsectivora, Paenungulata, Glires, Scrotifera, and Ferae.

This means that despite the progress made in recent years, there is still no consensus on quite a number of nodes within the placental tree, and morphological support for many of the molecular-based clades is often meager. Within Afrotheria, it is mainly the position of the aardvark (Tubulidentata[7]) and the interrelationships within the Paenungulata that account for differences among analyses. Many studies combine Tubulidentata with the Afroinsectivora into a clade Afroinsectiphilia (e.g., Meredith et al. 2011), but other studies favor a sister group relationship of Tubulidentata with Paenungulata (e.g., O'Leary et al. 2013a). Among the three extant highest-ranking paenungulate groups, elephants (Proboscidea) often are sister to the manatees and dugong (Sirenia), forming a clade called Tethytheria (e.g., O'Leary et al. 2013a, also in line with morphological similarities between elephants and sirenians such as the position of the mammary glands and the horizontal molar progression or "mesial drift"), but in a number of analyses, hyraxes (Hyracoidea) turn out to be sister to Proboscidea (e.g., Meredith et al. 2011).

Within Euarchontoglires, the sister group relationship of Rodentia and Lagomorpha is well established (Glires), but the (Eu-)Archonta are more problematic. Most importantly, the phylogenetic position of the tree shrews (Scandentia) is uncertain. This group was once included in the menotyphlan part of "Insectivora," and Simpson (1945) included it in the Lemuriformes within Primates (see above). Their being part of Euarchontoglires is beyond reasonable doubt, but although they are often included in (Eu-) Archonta, this is not definitive: Meredith et al. (2011) found a sister group relationship between Scandentia and Glires, and Tarver et al. (2016) one between Scandentia and all remaining Euarchontoglires. Esselstyn et al. (2017) favored a sister group relationship between Scandentia and Primatomorpha (Dermoptera + Primates), but could not rule out that Scandentia is sister to Glires. Finally, Scandentia might be the sister group of the colugos (Dermoptera) in a clade called Sundatheria (for a recent example, see Upham et al. 2019). This phylogenetic riddle therefore has implications for our own history – the question of the sister group of the primates. Is it Dermoptera (Primatomorpha hypothesis, see Fig. 4 and, e.g., Meredith et al. 2011 and Esselstyn et al. 2017) or Dermoptera and Scandentia combined (Sundatheria hypothesis, e.g., O'Leary et al. 2013a)? A definitive answer is lacking so far.

That the deepest split within Laurasiatheria is between (Eu-)Lipotyphla and the remaining taxa (Scrotifera) is well established. So is the sister group relationship between carnivorans and pangolins (combined into Ferae or Ostentoria[8]). However, except for Ferae, relationships within Scrotifera are among the most contentious for all higher taxa within Placentalia. Whether the two ungulate groups Perissodactyla and (Cet-) Artiodactyla form a clade (Euungulata[9]), for example, is not clear (but favored by, among others, Tarver et al. 2016 and Esselstyn et al. 2017). Neither is the position of Chiroptera. Tarver et al. (2016) and Esselstyn et al. (2017) present evidence for their being sister to all remaining Scrotifera (the latter clade is known as Fereuungulata). Alternatively, they might be sister to Euungulata (e.g., Meredith et al. 2011; O'Leary et al. 2013a) or part of a clade called Pegasoferae that comprises, apart from the bats, Perissodactyla and Ferae (carnivorans and pangolins) (Nishihara et al. 2006). Support for Pegasoferae has also been found by Zhang et al. (2013). Their analysis yielded Chiroptera as sister to Perissodactyla to the exclusion of Ferae. So did the phylogenomic analysis by Chen et al. (2017), which favored a sister group relationship of Chiroptera and Perissodactyla, which in turn were sister to Carnivora + Cetartiodactyla (Pholidota, unfortunately, were not included in the study).

A Temporal Framework for the Diversification of Mammals

Inferences of divergence times (age of nodes) depend on the first appearance of relevant groups in the fossil record and the estimation of evolutionary rates. The latter are mostly based on molecular clocks for DNA sequence data but can also comprise morphological clocks (e.g., Puttick et al. 2016). Different assumptions with respect to evolutionary rates, interpretations of fossils, and discoveries of new fossils as well as different methodologies applied to datasets will inevitably lead to different ages for the nodes in a phylogeny. For example, Archibald (2011, Fig. 3.1), based on the literature up to then and his own evaluations, gives an age of crown-group Mammalia of 150–180 mya and of 105 mya for the split between Eutheria and Metatheria (see also Fig. 2). However, since then, a new fossil has been described, *Juramaia sinensis*, that was interpreted as an early stem eutherian from the Jurassic some 160 mya (Luo et al. 2011). Before that, the oldest eutherian and metatherian (*Eomaia* and *Sinodelphys*, respectively) were considered to be ca. 125 mya. If *Juramaia* is indeed a eutherian and not a stem therian (the question seems to be still open, and the same holds for *Eomaia*, see, e.g., Averianov and Archibald 2016), the minimum age of the Eutheria/Metatheria split is 160 mya, also pushing the node of crown-group Mammalia back in time. According to a very recent study (Bi et al. 2018), *Sinodelphys* is not the oldest metatherian, but rather a eutherian, which means that the oldest metatherians are only 110 million years old, producing a 50-million-year ghost lineage for Metatheria. A detailed discussion and review of all the published divergence times for the major nodes within Mammalia is beyond the scope of this short overview. The comprehensive study by Tarver et al. (2016) estimates the root of Mammalia at between ca. 200 and 250 mya, the Eutheria/Metatheria split (i.e., the Theria node) at ca. 157–170 mya, the most basal Placentalia node at ca. 86–100 mya and the marsupial root at 49–104 mya. This is in line with other studies; for example, dos Reis et al. (2012) and Foley et al. (2016) published similar results. Even when taking into account that a recalculation triggered by a different fossil phylogeny (see the "Note Added in Proof" on p. 342 of Tarver et al. 2016) yielded a crown-group Mammalia node that was younger by 40 million years[10], this result is in better accordance with the recent fossil discoveries. The placental, or crown eutherian, node is of particular interest to zoologists as it refers to the most recent common ancestor of all living placental mammals. There is a growing consensus as to it being of Late Cretaceous age across a number of the most comprehensive molecular studies: 89.1 mya (dos Reis et al. 2012), 89.8 mya (Wu et al. 2017), 93 mya (Tarver et al. 2016), and 98.6 (Foley et al. 2016). Confidence intervals of these values are given for most studies, and they imply relatively narrow error margins.

In parallel with the debate about the phylogenetic relationships within Placentalia, there has been a discussion about the timeframe of the origin and diversification of the placental groups ("supraordinal" taxa and "orders"). This discussion has mainly been fueled by divergent results of molecular clock analyses and the discrepancy between molecular-based estimates and the first appearance of placental groups in the fossil record (e.g., Bininda-Emonds et al. 2007; Wible et al. 2007; Archibald 2011; Meredith et al. 2011; dos Reis et al. 2014; Puttick et al. 2016, and references therein). The core question is which, if any, nodes of the placental tree are older than the K-Pg boundary ca. 66 mya. Archibald and Deutschman (2001) coined new terms for three temporal evolutionary models of placental mammals: (i) the *explosive model*, according to which "most if not all interordinal origination and diversification as well as ordinal origination of extant placentals occurred within a very short interval of about 10 million years, mainly following the K/T boundary" (p. 111); (ii) the *long fuse model* that argues for a diversification of extant "orders" after the K-Pg boundary but allows for long stem lineages leading to "interordinal" ancestors that extend well back into the Late Cretaceous; and (iii) the *short fuse model* with origination and diversification of crown "orders" of placentals well before the K-Pg boundary[11]. A fourth, *soft explosive model* has recently been used by Phillips (2016, Phillips and Fruciano 2018). It differs from the normal "hard" explosive model "only by extending the earliest few placental superordinal divergences into the Cretaceous" (Phillips 2016, p. 547) and was criticized strongly by Springer et al. (2017) who, like many others, favor the long fuse model. One of the problems with the soft explosive model is that, according to critics, a large number of nodes are actually younger than minimum ages derived by the fossil record (dubbed "zombie lineages," in analogy to ghost lineages, by Springer et al. 2017; but see Phillips and Fruciano 2018 for a rebuttal). Foley et al.

(2016) give a concise discussion and critique of all four models.

Although in general a perfect match of molecular dating and the fossil record is not to be expected (and the discrepancy between them may be less pronounced than often thought, Cunningham et al. 2017), in the case of placental mammals, molecular estimates placing the origination and particularly the diversification of placentals in the Cretaceous are at odds with the repeated paleontological finding that there is not a single undisputed placental fossil older than the earliest Cenozoic (Asher et al. 2005; Wible et al. 2007; Goswami et al. 2011; Halliday et al. 2017). Not all molecular studies favor a Cretaceous origin of placental mammals, though, leading to the reverse pattern of paleontologists (and molecular biologists) defending a Mesozoic origin for Placentalia: O'Leary et al. (2013a, b) hypothesized that the most recent common ancestor of placentals lived after the K-Pg boundary, supporting the explosive model, but this has been rejected by other groups (e.g., Springer et al. 2013; dos Reis et al. 2014) and seems ultimately to have been based on the doubtful inference that the earliest known appearance of taxa in the Paleocene fossil record was in fact the actual time of origin for these groups. On the other hand, it has been argued that the morphology-based results may be flawed and that the fact that only stem eutherians predate the K-Pg boundary may be due to stemward slippage of fossils and problems with homoplasy (discussed in Foley et al. 2016). If that is true, the lack of Mesozoic placentals may be an artifact of the fossil record or its analysis. In a recent study, Halliday et al. (2019) found substantial decoupling of molecular and morphological rates of evolution with the latter being slow compared to the former during the initial divergence of placental mammals but then increasing during the origination of the extant "orders." As a consequence, early placental fossils may be difficult to distinguish from stem eutherians such that the "lack of definitive Cretaceous placental mammals [could also be due to] similarity among stem and early crown eutherians" (= placentals). Still, as it now stands, both the explosive and the short fuse model are not in accordance with either (most) molecular estimates or the lack of Mesozoic fossil placentals, making the long fuse model the most likely candidate (but see Phillips and Fruciano 2018). This model best combines molecular and paleontological data and has been supported by recent genomic divergence estimates published by dos Reis et al. (2012) and Tarver et al. (2016). Both these studies have placed the origins of extant placental "orders" at a 20-million-year window after the K-Pg boundary (roughly 45–65 mya). The only exceptions were xenarthrans and primates whose origins were dated to 56–77 mya and 65–73 mya, respectively, with means of 67 mya for xenarthrans and 69 mya for primates (for primates, see also dos Reis et al. 2018 who place the origin of primates at between 63.9 and 79.2 mya). It thus seems that paleontological and molecular results are converging toward a placental origin sometime in the Late Cretaceous with a diversification of the extant groups in the early Cenozoic.

How the K-Pg mass extinction impacted mammals is also not definitively known. For North America, Pires et al. (2018) found that only Metatheria showed increased extinction rates during the K-Pg, while Multituberculata mainly exhibited decreased origination rates; eutherians showed high turnover rates due to peaks in both origination and extinction rates. Not confining themselves to North America, Liu et al. (2017, p. 7282) conclude that "placental mammals underwent a continuous radiation across the KPg boundary without apparent interruption by the mass extinction, paralleling a genus-level radiation of multituberculates and ecomorphological diversification of both multituberculates and therians. These findings suggest that the KPg catastrophe evidently played a limited role in placental diversification, which, instead, was likely a delayed response to the slightly earlier radiation of angiosperms." This is in line with Benton (2010) who argued that the Cretaceous Terrestrial Revolution (KTR), when angiosperms became the dominant plant group between 125 and 80 mya, had a profound impact on mammalian diversification. Another, longstanding idea, namely, that newly available ecospace vacated by the K-Pg mass extinction was filled by early Paleocene mammal radiations, is also in line with this scenario, as pointed out by Liu et al. (2017, p. 7289) when they state that although placentals "radiated across the boundary with little interruption [. . .], the intraordinal diversification of placental mammals appears to have been a Cenozoic phenomenon, consistent with the traditional view that mammals benefited from numerous vacant niches following the catastrophe." The analysis of morphological rates of evolution leads Halliday et al. (2019, p. 7) to a very similar conclusion in that after the mass extinction, "release of ecological constraints and higher niche availability allowed morphological diversification [...]. With more avenues in the adaptive landscape through which to evolve (as a result of empty niches), the phenotypic result of any given mutation would be more likely to be beneficial *in some direction* (that is, into an empty niche)." If true, this would mean that after the End-Permian mass extinction had set the stage for the rise of the Archosauria (incl. the dinosaurs) throughout the Mesozoic by putting an end to the dominance of Paleozoic synapsids ("mammal-like reptiles" along the mammalian stem lineage), the K-Pg event again enabled the mammalian lineage to diversify, this time into the magnificent variety of its present 5,000–6,000 species – including, of course, our own.

Endnotes

1. When I talk of higher, high-ranking, lower-ranking, etc. taxa, this refers solely to the level of inclusiveness. Thus, Afrotheria is a higher-ranking taxon than, say, Proboscidea because it is more inclusive and its most recent ancestor lived further in the past. It has nothing to do with the outdated scala-naturae-based concept of higher vs. lower animals (see further below in the main text).
2. The exact number of species depends on whether species are delimited more or less inclusively, i.e., ultimately on whether one is more prone to lumping or splitting. For a recent, splitting-friendly census of ca. 6,500, see Burgin et al. (2018).
3. Terminology is sometimes imprecise. I prefer the following: Eutheria comprises stem Eutheria (a paraphyletic assemblage) and the monophyletic crown Eutheria. The latter are synonymous with Placentalia. Placentalia is thus defined as a crown group. Still, one often reads stem Placentalia when referring to stem Eutheria. This is misleading as stem Placentalia would in fact be outside Placentalia. When it is pointed out that there are no crown placental fossils from the Mesozoic, this is somewhat pleonastic as placentals are defined as a crown group. What is meant in the terminology preferred here is that stem eutherian fossils are known but crown eutherian (= placental) fossils are missing.
4. Linnaean categories are scientifically untenable, giving the same name to groups of different age and hierarchical level and thus creating an "apples-and-oranges" problem (see, e.g., Zachos 2011). Since they are used so frequently, however, it is almost impossible to refer to published work without mentioning them. They are used in inverted commas in this chapter to make it clear that they are artificial.
5. This has been confirmed by morphologically based phylogenies as well, and the key apomorphy of artiodactyls, the "double-pulley" astragalus, is known to have been present in fossil cetaceans (see Price et al. 2005; Geisler and Uhen 2005; Geisler et al. 2007; O'Leary and Gatesy 2008 and references therein). "Double-pulley" refers to the fact that the astragalus has both a proximal trochlea (for the tibia) and a distal one (for articulation with the navicular and cuboid). The clade comprising hippos and whales is called Whippomorpha or Cetancodonta.
6. There is disagreement about whether it should be Archonta or Euarchonta and Lipotyphla or Eulipotyphla, and the same applies to Artiodactyla and Cetartiodactyla. Archonta (introduced by Gregory 1910) originally not only comprised Scandentia,

Primates, and Dermoptera but also Macroscelidea and Chiroptera, and Euarchonta is supposed to denote that Chiroptera are excluded (elephant shrews have long been known not to be part of it). Similarly, Cetartiodactyla highlights the inclusion of Cetacea within the even-toed ungulates, and Eulipotyphla denotes the exclusion of the Afrosoricida, and some authors think this should be recognizable by the name (an analogous situation is the one of Sauropsida vs. Reptilia incl. Aves). Because Euarchontoglires is a new concept (there never was a name Archontoglires), the prefix Eu- is always used in this case. See Asher and Helgen (2010) for a discussion. These authors prefer the old names, and they also give a list of higher-level taxa within Placentalia and their first publication, which is an update of the list in Asher et al. (2009).

7. The recently extinct so-called Malagasy aardvark *Plesiorycteropus* from Madagascar (sometimes classified in its own "order" Bibymalagasia) is in fact more likely a giant tenrec and thus part of Afrosoricida (Buckley 2013 and references therein).
8. When only considering extant taxa, Ferae and Ostentoria denote the same clade but Ferae usually also includes a number of fossil groups such that Ostentoria is the extant crown group within the more inclusive Ferae.
9. The prefix Eu- in Euungulata denotes the fact that groups formerly comprised by Ungulata are no longer included, namely, Tubulidentata, Hyracoidea, Proboscidea, and Sirenia (see Asher and Helgen 2010).
10. The age of all other nodes did not change markedly.
11. Archibald and Deutschman (2001) take the term fuse from Cooper and Fortey (1998) who called "phylogenetic fuse" the time between the origination and the often seemingly explosive adaptive radiation of a group, an idea the authors trace back to observations made by George Gaylord Simpson.

Acknowledgments I would like to thank Vera Weisbecker from the University of Queensland in Brisbane, Australia, for a constructive review of an earlier version of this chapter and Kriemhild Repp from the Natural History Museum Vienna for drawing the phylogenetic trees.

References

Angielczyk KD, Kammerer CF (2018) Non-mammalian synapsids: the deep roots of the mammalian family tree. In: Zachos FE, Asher RJ (eds) Mammalian evolution, diversity and systematics. Handbook of zoology, Mammalia. de Gruyter, Berlin, pp 117–198

Archibald JD (2011) Extinction and radiation. How the fall of dinosaurs led to the rise of mammals. The Johns Hopkins University Press, Baltimore

Archibald JD, Deutschman DH (2001) Quantitative analysis of the timing of the origin and diversification of extant placental orders. J Mamm Evol 8:107–124

Asher RJ (2018) Diversity and relationships within crown Mammalia. In: Zachos FE, Asher RJ (eds) Mammalian evolution, diversity and systematics. Handbook of zoology, Mammalia. de Gruyter, Berlin, pp 301–352

Asher RJ, Helgen KM (2010) Nomenclature and placental mammal phylogeny. BMC Evol Biol 10:102

Asher RJ, Horovitz I, Sánchez-Villagra MR (2004) First combined cladistics analysis of marsupial mammal interrelationships. Mol Phylogenet Evol 33:240–250

Asher RJ, Meng J, Wible JR, McKenna MC, Rougier GW, Dashzeveg D, Novacek MJ (2005) Stem Lagomorpha and the antiquity of Glires. Science 307:1091–1094

Asher RJ, Bennett N, Lehmann T (2009) The new framework for understanding placental mammal evolution. BioEssays 31:853–864

Averianov AO, Archibald D (2016) New evidence on the stem placental mammal *Paranyctoides* from the Upper Cretaceous of Uzbekistan. Palaeontol Pol 67:25–33

Baum DA, Smith SD, Donovan SSS (2005) The tree-thinking challenge. Science 310:979–980

Beck RMD, Travouillon KJ, Aplin KP, Godthelp H, Archer M (2014) The osteology and systematics of the enigmatic Australian Oligo-Miocene metatherian Yalkaparidon (Yalkaparidontidae; Yalkaparidontia; ? Australidelphia; Marsupialia). J Mamm Evol 21:127–172

Begun DR (1992) *Dryopithecus crusafonti* sp. nov., a new Miocene hominoid species from Can Ponsic (northeastern Spain). Am J Phys Anthropol 87:291–309

Benton MJ (2010) The origins of modern biodiversity on land. Philos Trans R Soc B 365:3667–3679

Bi S, Zheng X, Wang X, Cignetti NE, Yang S, Wible JR (2018) An Early Cretaceous eutherian and the placental-marsupial dichotomy. Nature 558:390–395

Bininda-Emonds ORP, Cardillo M, Jones KE, MacPhee RDE, Beck RMD, Grenyer R, Price SA, Vos RA, Gittleman JL, Purvis A (2007) The delayed rise of present-day mammals. Nature 446:507–512

Böhme M, Spassov N, Fuss J, Tröscher A, Deane AS, Prieto J, Kirscher U, Lechner T, Begun DR, (2019) A new Miocene ape and locomotion in the ancestor of great apes and humans. Nature 575 (7783):489–493

Bond M, Tejedor MF, Campbell KE Jr, Chornogubsky L, Novo N, Goin F (2015) Eocene primates of South America and the African origins of New World monkeys. Nature 520:538–541

Brace S, Thomas JA, Dalén L, Burger J, MacPhee RDE, Barnes I, Turvey ST (2016) Evolutionary history of the Nesophontidae, the last unplaced recent mammal family. Mol Biol Evol 33:3095–3103

Buckley M (2013) A molecular phylogeny of *Plesiorycteropus* reassigns the extinct mammalian order 'Bibymalagasia'. PLoS One 8:e59614

Burgin CJ, Colella JP, Kahn PL, Upham NS (2018) How many species of mammals are there? J Mammal 99:1–14

Chen M-Y, Liang D, Zhang P (2017) Phylogenomic resolution of the phylogeny of Laurasiatherian mammals: exploring phylogenetic signals within coding and noncoding sequences. Genome Biol Evol 9:1998–2012

Cooper A, Fortey R (1998) Evolutionary explosions and the phylogenetic fuse. Trends Ecol Evol 13:151–156

Cunningham JA, Liu AG, Bengtson S, Donoghue PCJ (2017) The origin of animals: can molecular clocks and the fossil record be reconciled? BioEssays 39:1–12

dos Reis M, Inoue J, Hasegawa M, Asher RJ, Donoghue PCJ, Yang Z (2012) Phylogenomic datasets provide both precision and accuracy in estimating the timescale of placental mammal phylogeny. Proc R Soc B 279:3491–3500

dos Reis M, Donoghue PCJ, Yang Z (2014) Neither phylogenomic nor palaeontological data support a Palaeocene origin of placental mammals. Biol Lett 10:20131003

dos Reis M, Gunnell GF, Barba-Montoya J, Wilkins A, Yang Z, Yoder AD (2018) Using phylogenomic data to explore the effects of relaxed clocks and calibration strategies on divergence time estimation: primates as a test case. Syst Biol 67:594–615

Duchêne DA, Bragg JG, Duchêne S, Neaves LE, Potter S, Moritz C, Johnson RN, Ho SYW, Eldridge MDB (2018) Analysis of phylogenomic tree space resolves relationships among marsupial families. Syst Biol 67:400–412

Eldridge MDB, Beck RMD, Croft DA, Travouillon KJ, Fox BJ, (2019) An emerging consensus in the evolution, phylogeny, and systematics of marsupials and their fossil relatives (Metatheria). Journal of Mammalogy 100(3):802–837

Esselstyn JA, Oliveros CH, Swanson MT, Faircloth BC (2017) Investigating difficult nodes in the placental mammal tree with expanded taxon sampling and thousands of ultraconserved elements. Genome Biol Evol 9:2308–2321

Foley NM, Springer MS, Teeling EC (2016) Mammal madness: is the mammal tree of life not yet resolved? Philos Trans R Soc B 371:20150140

Gaudin TJ, Emry RJ, Wible JR (2009) The phylogeny of living and extinct pangolins (Mammalia, Pholidota) and associated taxa: a morphology based analysis. J Mamm Evol 16:235–305

Geisler JH, Uhen MD (2005) Phylogenetic relationships of extinct cetartiodactyls: results of simultaneous analyses of molecular, morphological, and stratigraphic data. J Mamm Evol 12:145–160

Geisler JH, Theodor JM, Uhen MD, Foss SE (2007) Phylogenetic relationships of cetaceans to terrestrial artiodactyls. In: Prothero DR, Foss SE (eds) The evolution of artiodactyls. The Johns Hopkins University Press, Baltimore, pp 19–31

Goswami A, Prasad GV, Upchurch P, Boyer DM, Seiffert ER, Verma O, Gheerbrant E, Flynn JJ (2011) A radiation of arboreal basal eutherian mammals beginning in the Late Cretaceous of India. Proc Natl Acad Sci 108:16333–16338

Gregory WK (1910) The orders of mammals. Bull Am Mus Nat Hist 27:1–524

Gregory WK (1947) The monotremes and the palimpsest theory. Bull Am Mus Nat Hist 88:1–52

Halliday TJD, Upchurch P, Goswami A (2017) Resolving the relationships of Paleocene placental mammals. Biol Rev 92:521–550

Halliday TJD, Mario dos Reis, Tamuri AU, Ferguson-Gow H, Yang Z, Goswami A (2019) Rapid morphological evolution in placental mammals post-dates the origin of the crown group. Proceedings of the Royal Society B: Biological Sciences 286(1898):20182418

Horovitz I, Sánchez-Villagra MR (2003) A morphological analysis of marsupial mammal higher-level phylogenetic relationships. Cladistics 19:181–212

Janke A, Magnell O, Wieczorek G, Westerman M, Arnason U (2002) Phylogenetic analysis of 18S rRNA and the mitochondrial genomes of the wombat, *Vombatus ursinus*, and the spiny anteater, *Tachyglossus aculeatus*: increased support for the Marsupionta hypothesis. J Mol Evol 54:71–80

Kullberg M, Hallström BM, Arnason U, Janke A (2008) Phylogenetic analysis of 1.5 Mpb and platypus EST data refute the Marsupionta hypothesis and unequivocally support Monotremata as sister group to Marsupialia/Placentalia. Zool Scr 37:115–127

Liu L, Zhan J, Rheindt FE, Lei F, Qu Y, Wang Y, Zhang Y, Sullivan C, Nie W, Wang J, Yang F, Chen J, Edwards SV, Meng J, Wu S (2017) Genomic evidence reveals a radiation of placental mammals uninterrupted by the KPg boundary. Proc Natl Acad Sci U S A 114: E7282–E7290

Luo Z-X, Yuan C-Y, Meng Q-J, Ji Q (2011) A Jurassic eutherian mammal and divergence of marsupials and placentals. Nature 476:442–445

Madsen O, Scally M, Douady CJ, Kao DJ, DeBry RW, Adkins R, Amrine HM, Stanhope MJ, de Jong WW, Springer MS (2001) Parallel adaptive radiations in two major clades of placental mammals. Nature 409:610–614

Martin T (2018) Mesozoic mammals. In: Zachos FE, Asher RJ (eds) Mammalian evolution, diversity and systematics. Handbook of zoology, Mammalia. de Gruyter, Berlin, pp 199–299

May-Collado LJ, Kilpatrick CW, Agnarsson I (2015) Mammals from "down under": a multi-gene species-level phylogeny of marsupial mammals (Mammalia, Metatheria). PeerJ 3:e805

Meredith RW, Westerman M, Case JA, Springer MS (2008) A phylogeny and timescale for marsupial evolution based on sequences for five nuclear genes. J Mamm Evol 15:1–36

Meredith RW, Janečka JE, Gatesy J, Ryder OA, Fisher CA, Teeling EC, Goodbla A, Eizirik E, Simão TLL, Stadler T, Rabosky DL, Honeycutt RL, Flynn JL, Ingram CM, Steiner C, Williams TL, Robinson TJ, Burk-Herrick A, Westerman M, Ayoub NA, Springer MS, Murphy WJ (2011) Impacts of the Cretaceous terrestrial revolution and KPg extinction on mammal diversification. Science 334:521–524

Mitchell KJ, Pratt RC, Watson LN, Gibb GC, Llamas B, Kasper M, Edson J, Hopwood B, Male D, Armstrong KN, Meyer M, Hofreiter M, Austin J, Donnellan SC, Lee MSY, Phillips MJ, Cooper A (2014) Molecular phylogeny, biogeography, and habitat preference evolution of marsupials. Mol Biol Evol 31:2322–2330

Morgan CC, Foster PG, Webb AE, Pisani D, McInerney JO, O'Connell MJ (2013) Heterogenous models place the root of the placental mammal phylogeny. Mol Biol Evol 30:2145–2156

Moyà-Solà S, Köhler M, Alba DM, Casanovas-Vilar I, Galindo J (2004) *Pierolapithecus catalaunicus*, a new Middle Miocene great ape from Spain. Science 306:1339–1344

Murphy WJ, Eizirik E, Johnson WE, Zhang YP, Ryder OA, O'Brien SJ (2001a) Molecular phylogenetics and the origins of placental mammals. Nature 409:614–618

Murphy WJ, Eizirik E, O'Brien SJ, Madsen O, Scally M, Douady CJ, Teeling E, Ryder OA, Stanhope MJ, de Jong WW, Springer MS (2001b) Resolution of the early placental mammal radiation using Bayesian phylogenetics. Science 294:2348–2351

Murphy WJ, Pringle TH, Crider TA, Springer MS, Miller W (2007) Using genomic data to unravel the root of the placental mammal phylogeny. Genome Res 17:413–421

Nilsson MA, Arnason U, Spencer PBS, Janke A (2004) Marsupial relationships and a timeline for marsupial radiation in South Gondwana. Gene 340:189–196

Nishihara H, Hasegawa M, Okada N (2006) Pegasoferae, an unexpected mammalian clade revealed by tracking ancient retroposon insertions. Proc Natl Acad Sci 103:9929–9934

Nishihara H, Maruyama S, Okada N (2009) Retroposon analysis and recent geological data suggest near-simultaneous divergence of the three superorders of mammals. Proc Natl Acad Sci 106:5235–5240

Novacek MJ (1992) Mammalian phylogeny: shaking the tree. Nature 356:121–125

O'Leary MA, Gatesy J (2008) Impact of increased character sampling on the phylogeny of Cetartiodactyla (Mammalia): combined analysis including fossils. Cladistics 24:397–442

O'Leary MA, Bloch JI, Flynn JJ, Gaudin TJ, Giallombardo A, Giannini NP, Goldberg SL, Kraatz BP, Luo Z-X, Meng J, Ni X, Novacek MJ, Perini FA, Randall ZS, Rougier GW, Sargis EJ, Silcox MT, Simmons NB, Spaulding M, Velazco PM, Weksler M, Wible JR, Cirranello AL (2013a) The placental mammal ancestor and the post-K-Pg radiation of placentals. Science 339:662–667

O'Leary MA, Bloch JI, Flynn JJ, Gaudin TJ, Giallombardo A, Giannini NP, Goldberg SL, Kraatz BP, Luo Z-X, Meng J, Ni X, Novacek MJ, Perini FA, Randall ZS, Rougier GW, Sargis EJ, Silcox MT, Simmons NB, Spaulding M, Velazco PM, Weksler M, Wible JR, Cirranello AL (2013b) Response to comment on "The placental mammal ancestor and the post-K-Pg radiation of placentals". Science 341:613

Omland KE, Cook LG, Crisp MD (2008) Tree thinking for all biology: the problem with reading phylogenies as ladders of progress. BioEssays 30:854–867

Phillips MJ (2016) Geomolecular dating and the origin of placental mammals. Syst Biol 65:546–557

Phillips MJ, Fruciano C (2018) The soft explosive model of placental mammal evolution. BMC Evol Biol 18:104

Pires MM, Rankin BD, Silvestro D, Quental TB (2018) Diversification dynamics of mammalian clades during the K-Pg mass extinction. Biol Lett 14:20180458

Price SA, Bininda-Emonds ORP, Gittleman JL (2005) A complete phylogeny of the whales, dolphins and even-toed hoofed mammals (Cetartiodactyla). Biol Rev 80:445–473

Puttick MN, Thomas GH, Benton MJ (2016) Dating placentalia: morphological clocks fail to close the molecular fossil gap. Evolution 70:873–886

Rigato E, Minelli A (2013) The great chain of being is still here. Evol Educ Outreach 6:18

Romiguier J, Ranwez V, Delsuc F, Galtier N, Douzery EJP (2013) Less is more in mammalian phylogenomics: AT-rich genes minimize tree conflicts and unravel the root of placental mammals. Mol Biol Evol 30:2134–2144

Rose KD (1999) *Eurotamandua* and Palaeanodonta: convergent or related? Paläontol Z 73:395–401

Sánchez-Villagra MR (2013) Why are there fewer marsupials than placentals? On the relevance of geography and physiology to evolutionary patterns of mammalian diversity and disparity. J Mamm Evol 20:279–290

Sánchez-Villagra MR, Narita Y, Kuratani S (2007) Thoracolumbar vertebral number: the first skeletal synapomorphy for afrotherian mammals. Syst Biodivers 5:1–7

Scally M, Madsen O, Douady CJ, de Jong WW, Stanhope MJ, Springer MS (2001) Molecular

evidence for the major clades of placental mammals. J Mamm Evol 8:239–277
Scotese CR (2001) Atlas of earth history. University of Texas at Arlington. Department of Geology. PALEOMAP Project
Simpson GG (1945) The principles of classification and a classification of mammals. Bull Am Mus Nat Hist 85.:xvi +:1–350
Springer MS, Cleven GC, Madsen O, de Jong WW, Waddell VG, Amrine HM, Stanhope MJ (1997) Endemic African mammals shake the phylogenetic tree. Nature 388:61–64
Springer MS, Stanhope MJ, Madsen O, de Jong WW (2004) Molecules consolidate the placental mammal tree. Trends Ecol Evol 19:430–438
Springer MS, Meredith RW, Teeling EC, Murphy WJ (2013) Technical comment on "The placental mammal ancestor and the post-K-Pg radiation of placentals". Science 341:613
Springer MS, Emerling CA, Meredith RW, Janečka JE, Eizirik E, Murphy WJ (2017) Waking the undead: implications of a soft explosive model for the timing of placental mammal diversification. Mol Phylogenet Evol 106:86–102
Stanhope MJ, Waddell VG, Madsen O, de Jong WW, Hedges SB, Cleven GC, Kao D, Springer MS (1998) Molecular evidence for multiple origins of Insectivora and for a new order of endemic African insectivore mammals. Proc Natl Acad Sci 95:9967–9972
Storch G (1981) *Eurotamandua joresi*, ein Myrmecophagide aus dem Eozän der "Grube Messel" bei Darmstadt (Mammalia, Xenarthra). Senckenb Lethaea 61:247–289
Tarver JE, dos Reis M, Mirarab S, Moran RJ, Parker S, O'Reilly JE, King BL, O'Connell MJ, Asher RJ, Warnow T, Peterson KJ, Donoghue PCJ, Pisani D (2016) The interrelationships of placental mammals and the limits of phylogenetic inference. Genome Biol Evol 8:330–344
Teeling EC, Hedges SB (2013) Making the impossible possible: rooting the tree of placental mammals. Mol Biol Evol 30:1999–2000
Upham NS, Esselstyn JA, Jetz W (2019) Inferring the mammal tree: Species-level sets of phylogenies for questions in ecology, evolution, and conservation. PLOS Biology 17(12):e3000494
Weisbecker V (2015) Are monotremes primitive and marsupials inferior? In: Klieve A, Hogan L, Johnston S, Murray P (eds) Marsupials and monotremes. Nature's enigmatic mammals. Nova Science Publisher, New York, pp 397–411
Wible JR, Rougier GW, Novacek MJ, Asher RJ (2007) Cretaceous eutherians and Laurasian origin for placental mammals near the K/T boundary. Nature 447:1003–1006
Wildman DE, Uddin M, Opazo JC, Liu G, Lefort V, Guindon S, Gascuel O, Grossman LI, Romero R, Goodman M (2007) Genomics, biogeography, and the diversification of placental mammals. Proc Natl Acad Sci 104:14395–14400
Wilson GP, Ekdale EG, Hoganson JW, Calede JJ, Linden AV (2016) A large carnivorous mammal from the Late Cretaceous and the North American origin of marsupials. Nat Commun 7:13734
Wu J, Yonezawa T, Kishino H (2017) Rates of molecular evolution suggest natural history of life history traits and a post-K-Pg nocturnal bottleneck of placentals. Curr Biol 27:3025–3033
Zachos FE (2011) Linnean ranks, temporal banding and time-clipping: why not slaughter the sacred cow? Biol J Linn Soc 103:732–734
Zachos FE (2016) Tree thinking and species delimitation: guidelines for taxonomy and phylogenetic terminology. Mamm Biol 81:185–188
Zhang G, Cowled C, Shi Z, Huang Z, Bishop-Lilly KA, Fang X, Wynne JW, Xiong Z, Baker ML, Zhao W, Tachedjian M, Zhu Y, Zhou P, Jiang X, Ng J, Yang L, Wu L, Xiao J, Feng Y, Chen Y, Sun X, Zhang Y, Marsh GA, Crameri G, Broder CC, Frey KG, Wang L-F, Wang J (2013) Comparative analysis of bat genomes provides insight into the evolution of flight and immunity. Science 339:456–460

Mammal Habitats in Europe: Geology, Vegetation, and Climate

4

Karl-Georg Bernhardt

Contents

K.-G. Bernhardt (✉)
Department of Integrative Biology and Biodiversity Research, Institute of Botany, University of Life Sciences and Natural Resources, Vienna (BOKU), Vienna, Austria
e-mail: karl-georg.bernhardt@boku.ac.at

K. Hackländer, F. E. Zachos (eds.), *Mammals of Europe - Past, Present, and Future*, Handbook of the Mammals of Europe, https://doi.org/10.1007/978-3-030-00281-7_2

Abstract

This chapter starts with the physiogeographical structure of Europe, its borders, areas, and locations. Some parts of Macaronesia (Madeira, Canary Islands, Azorean islands) are added. The geology, orography, and soils are described in the summary. A detailed physiogeographical organization describes ten areas of Northern Europe, Western and Central Europe, Southern Europe, Eastern Europe, and Macaronesia. They each are subdivided in different physiogeographical regions and provinces. Geographical location, landmass distribution, and topography determine the climatic patterns of Europe. Five main thermal zones can be distinguished from north to south: arctic, boreal, temperate, submeridional, and meridional. The climatic classification of Europe is the basis for the phytogeographical division of Europe. It includes floristic zones and regional subdivisions. At the end the Late-Glacial and Holocene vegetation history of Europe is summarized.

Keywords

Geology · Climate · Biogeography · Vegetation history · Landscape development

Introduction

Landscapes, habitats, and organisms in Europe developed as a result of geological and climatic changes. Geological processes and ecological situations up to the last ice age have formed European landscapes. A decrease of species during the ice age changed into an increase of species due to rewarming of the climate and anthropogenous activity forming a cultivated landscape.

All these forming processes differ over Europe by influence of increasing continental climate, degree of latitude, and elevation and form a couple of different climatic and geological situations. These differences can be monitored by vegetation which can be arranged by transects from North to South or East to West. During vegetation history, a lot of different areal types were formed. The floristic composition might be important for the differentiation of vegetation types but also influence of humans and animals. Seeing vegetation types as biocoenosis, we are talking about habitats. Therefore, the geologic, climatic, and cultural development of Europe is important for building mammal areas with their habitats.

Following the political situation, we add parts of Macaronesia (Canary Islands, the Azores, and Madeira archipelago) to Europe.

Physiographical Structuring of Europe

The Borders, Area, and Location of Europe

Europe is part of the Eurasian supercontinent. At the end of the Tertiary Period, its current form developed. Its geographical location and paleogeographic development have been most influencing parameters in forming Europe's highly pronounced physical differentiation and thus have affected profoundly biogeographical relationships.

On three sides, Europe's geographical borders are clearly defined: in the west by the Atlantic Ocean; in the north by the Norwegian Sea, the Barents Sea, and the Arctic Ocean; and in the south by the Mediterranean Sea and the Black Sea. In the east, the eastern slope of the Urals has commonly been set as the boundary to Asia due to paleogeographic, structural, and climatic aspects. It continues to the south of the Ural River and reaches the western shore of the Caspian Sea. The Great Caucasus is generally included in Europe today (Kondracki and Schlüter 2003).

Europe covers approximately 10 million km^2, excluding the Macaronesian region. About 0.74 million km^2 are located on islands genetically associated with the European mainland. The largest are Great Britain (224,000 km^2), Ireland (84,000 km^2), Sicily (26,000 km^2), Sardinia (24,000 km^2), and the genetically independent volcanic island of Iceland (103,000 km^2). The major peninsulas are the Scandinavian (824,000 km^2), Iberian (587,000 km^2), Apennine (149,000 km^2), and Balkan peninsulas (about 500,000 km^2). Islands and peninsulas cover well over one third of the total area of Europe. The diverse coastline has an overall length of 37,200 km (Kondracki and Schlüter 2003).

For our handbook, we include Macaronesia, a collection of four archipelagos in the North Atlantic Ocean off the coast of the continents of Europe and Africa. Apart from the Azores which are considered mainly as part of Europe, the islands of Macaronesia are geographically closer to Africa. Madeira and the Canary Islands are part of our distribution study, but not Cape Verde. The Azorean islands cover an area of 2,333 km^2, Madeira Islands 801 km^2, and Canary Islands 7,493 km^2 (Afonso 1988; Sziemer 2000).

From south to north, the mainland extends over 3,900 km, latitude between 36° and 71° N, including the islands over 5,300 km and about 48° of latitude. From west to east, it spreads about 5,000 km, that is, 78° of longitude. The northernmost point of Europe lies at 82° N on the Arctic islands of Svalbard and Franz Joseph Land and on the mainland at North Cape (71° 16' N). The furthest point in the southwest is Cape Marroqui in Spain, which reaches as far as 35° 58' N. The Mediterranean island of Crete lies about 1° further to the south. The westernmost point is at 9° 27' W at Cabo da Roca in Portugal, although the island of Ireland projects 1° further to the west (Iceland is not considered here since it is in fact an isolated island in the North Atlantic). The easternmost point of Europe is the mouth of the Bajdarata River that flows into the Kara Sea at 68° 14′ E.

The Azorean islands are a group of three archipelagos composed of nine inhabited islands in the Atlantic Ocean between 36° 55′ and 39° 43′ northern latitude and between 24° 46′ and 31° 16′ western longitude. The shortest distance to the European (Portuguese) coast is about 1,300 km. The eastern archipelago consists of the islands of Santa Maria, Sao Miguel, and the uninhabited Formigas Reef. The central archipelago consists of the islands of Terceira, Sao Jorge, Graciosa, Pico, and Faial. The two smaller islands, Ilha das Flores and Corvo, make up the western archipelago. The distance between the eastern- and westernmost islands, Santa Maria and Ilha das Flores, respectively, is about 600 km (Schäfer 2002; Dias et al. 2005).

Madeira is a Portuguese island and is the largest of the Madeira archipelago. The location is 32° 39'N and 16° 54'W. It covers an area of 741 km^2, 57 km in length (from Ponte de São Lourenço to Ponta do Pargo), while approximately 22 km at its widest point (from Ponte da Cruz to Ponte São Jorge), with a coastline of 150 km (Wirthmann 1970).

The Canary Islands are located in the Atlantic Ocean, about 100 to 500 km west of Morocco. The main islands are (from largest to smallest) Tenerife, Fuerteventura, Gran Canaria, Lanzarote, La Palma, La Gomera, and El Hierro. Included are also a number of smaller islands and islets: La Graciosa, Alegranza, Isla los Lobos, Montaña Clara, Roque del Oeste, and Roque del Este. Coordinates for the location are 27° 38′ to 29° 25' N and 13° 25′ to 18° 10' W. El Hierro, the westernmost island, covers 268.71 km^2, making it the smallest of the major islands. Fuerteventura, with a surface of 1,660 km^2, the most ancient of the islands, is the one that is eroded the most. Gran Canaria's surface area is 1,560 km^2. In the center of the island lie Roque Nublo (1,813 m) and Pico de las Nieves ("Peak of Snow" 1,949 m). La Gomera has an area of 369.76 km^2. Geologically, it is one of the oldest islands of the archipelago. Lanzarote is the easternmost island and also one of the most ancient of the archipelago, and it has shown evidence of recent volcanic activity. It has a surface of 8,455.94 km^2. The Chinijo Archipelago includes the islands La Graciosa, Alegranza, Montaña Clara, Roque del Este, and Roque del

Oeste. It has a surface of 40.8 km^2. With 29 km^2, La Graciosa is the smallest inhabited island of the Canaries and the major island of the Chinijo Archipelago (Acosta et al. 2005).

La Palma, covering an area of 708.32 km^2, shows no recent signs of volcanic activity, even though the volcano Teneguía last entered into eruption in 1971. In addition, it is the second-highest island of the Canaries, with the Roque de los Muchachos (2,423 m) as highest point. Tenerife is, with its area of 2,034 km^2 the most extensive island of the Canary Islands. The Teide, with 3,718 m is the highest peak of Spain.

Geology, Orography, and Soils

Paleogeographic development, especially tectonic processes in the Tertiary and repeated glaciation in the Quaternary, formed Europe's present-day surface. This diverse relief is the crucial factor for its physiographic structure (following Kondracki and Schlüter 2003).

The oldest part of the European mainland is the Precambrian basement, which is exposed as the Baltic Shield (Fennoscandia) in the north and the Ukrainian Shield in the southeast. Between these areas in the east European lowlands, these ancient rocks are overlain with Paleozoic and Mesozoic strata. This Palaeo-Europe was separated by a fault zone from Meso- and Neo-Europe. Meso-Europe arose – similar to the Urals – at the end of the Paleozoic during the Hercynian folding. During the Mesozoic these structures were levelled and covered by a series of marine and continental sediments.

Neo-Europe began to develop in the Mesozoic and attained its final shape in the Tertiary, when the folded structures of the high mountain regions (Sierra Nevada, Pyrenees, Alps, Apennines, Dinaric, Pindos and Balkan Mountains, Carpathian Mountains, and also the Caucasus) were uplifted. During this vertical movement, the horizontally folded rock formations were uplifted, while the Mediterranean and intramontane basins were lowered (Kondracki and Schlüter 2003).

wThe last major phase of relief formation was the repeated Quaternary glaciation – apart from erosion persisting to the present day. The main areas of the glaciation were Fennoscandia, the British Isles, and the Alps (Fig. 1). From the north, massive glaciers moved far into the Eastern, Central, and Western European plains and reached the Harz and Sudeten mountains and even the Western Carpathian Mountains. During the last glacial period, an inland ice sheet covered the Baltic Sea and the mainland to Jütland, the lower courses of the Elbe and the Oder valley, the middle Weichsel and Memel valleys, as well as the Duena valley (Andersen and Borns 1994).

In the cold periglacial climate, permafrost soils were widespread even outside of the glaciated regions. Especially in winter windblown dust material was deposited, often as thick loess, on the peripheral tundra and cold steppes surfaces. At the end of the last glaciation, glaciofluvial and fluvial sands were blown by wind to form dunes in areas that were still forest-free. Forests established in these areas only at the beginning of the Holocene.

In Fennoscandia and the other centers of glaciation, ancient rock formations were exposed by the retreating ice, while in the adjoining peripheral areas around the Baltic Sea, glacial till was deposited as moraines and glaciofluvial and glaciolimnic sediments (sands, gravels, clays). These deposits can be up to several hundred meters thick and rise up to 300 m above sea level.

Due to the postglacial warming, glaciers melted which caused the general rise of the mean sea level, the flooding of low mainland areas, the formation of the Baltic and North Seas, and the separation of the British Isles from the European mainland. This inland advance of coastline, or "transgression," was simultaneous to the isostatic elevation of areas earlier covered by massive glaciers (particularly in Fennoscandia), a process that continues today. The upward thrust of these areas is causing a recession particularly of the northern coast of the Baltic Sea, while on its southwestern shores as well as on the southern coasts of the North Sea, by contrast, the coastline is advancing inland in the course of further transgression (Kondracki and Schlüter 2003).

77The orography of Europe is quite diverse. Lowlands predominate the broad eastern half of

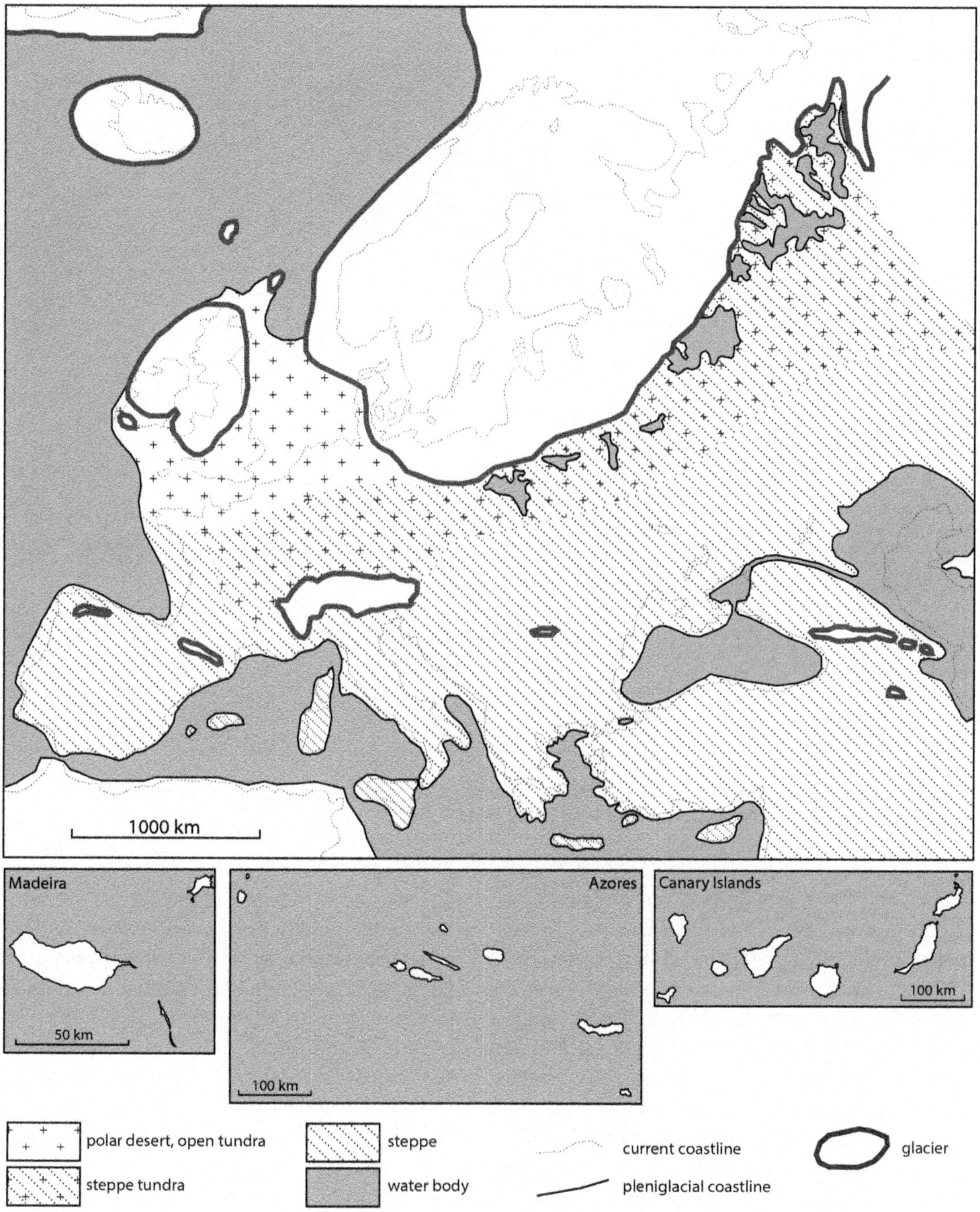

Fig. 1 Vegetation of Europe during the last glaciation around 20,000 BP (maximum expansion of ice during the last ice age). Map recreated and modified, area boundaries following Andersen and Borns (1994) and Lang (1994)

the continent which results in the mean elevation of Europe being only about 340 m above sea level (compared to 960 m in Asia) (Fig. 2). In contrast prominent mountain ranges dominate the southern half of Europe. These include the Alps (Mont Blanc 4,807 m), the Pyrenees (Pico de Aneto 3,404 m), the Apennines (Corno Grande 2,914 m), the Balkan Mountains (Botev

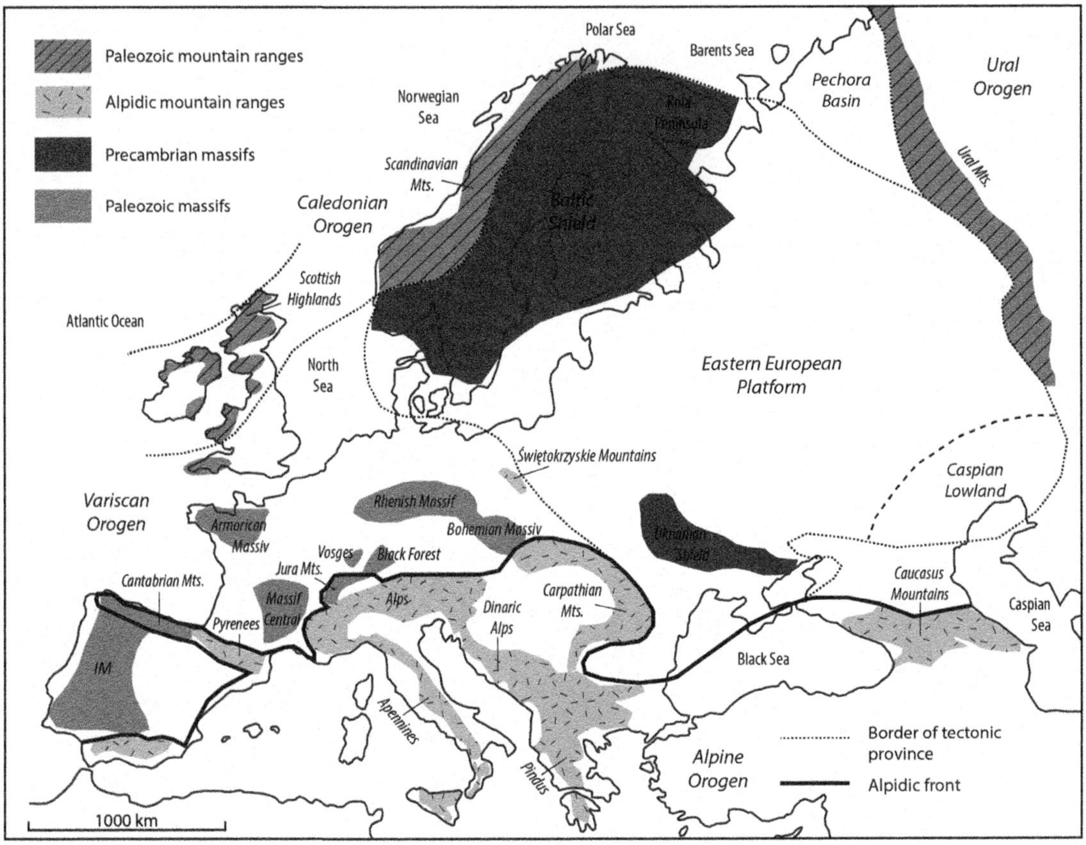

Fig. 2 High and low mountain ranges: Geological significance of important topographical features of Europe. Map recreated and modified, area boundaries following Park (2015)

2,376 m), the Carpathian Arch (High Tatras 2,655 m), and the Great Caucasus with seven peaks higher than 5,000 m (Elbrus 5,642 m). Hilly uplands extend to the west and north of the Alps (in the west, the French Massif Central (1886 m) with the Cevennes (1,699 m) and the Jura (1,718 m); in the northwest, the Vosges (1,424 m) and the Black Forest (1,493 m); and in the northeast, the Bavarian and Bohemian Forests (1,457 m) followed by the Hercynian mountain ranges, of which the Harz (1,142 m) extends furthest into the north German lowlands), while the ridge of the Erzgebirge (1,244 m) continues on to the east to the adjacent Sudeten mountains (1,603 m). The south Central European Uplands have counterparts in the Scandinavian highlands of Norway (2,470 m) as well as the Urals (1,894 m), of which the latter represent the northeastern boundary of the European mainland (Kondracki and Schlüter 2003).

In Macaronesia, several volcanic peaks exist. The highest peaks in the Azorean islands are Montanha do Pico (Pico Island, 2,351 m), Pico da Vara (São Miguel, 1,103 m), and Pico da Esperança (São Jorge, 1,053 m). The Azores are of volcanic origin without any parts of continental crust. They are located close to the Mid-Atlantic Ridge, a slow-moving sea floor-spreading zone, reaching from Jan Mayen and Iceland in the north to Tristan da Cunha in the south (Carracedo et al. 1998).

In the Azores region, the Eurasian, the African, and the American plates are drifting away from each other, causing constant tectonic disturbances. Magma reaches the upper parts of the crust and escapes, forming volcanoes. Today, 12 active volcanoes, 5 of them submarine, can be found in the Azores. Examinations of rocks and fossils revealed important differences in age between the islands of the area. Santa Maria, the easternmost island, developed probably in the Miocene about 5 to 14 million years ago. The northeastern parts of São Miguel are said to be 4 million years old, whereas the western parts seem to have existed for less than 2 million years. The lavas of Faial, São Jorge, and Graciosa were produced less than 750,000 years ago. With its 40,000 years, Pico, the highest mountain of Portugal, is the youngest island of the group. In general, the Azores are much younger than most of the other Atlantic Islands: the Madeira group is said to be 60 to 70 million years old, and the age of the Canary Islands ranges between 3 and 25 million years (Schäfer 2002).

Madeira is at the top of a massive shield volcano that rises about 6 km from the floor of the Atlantic Ocean, on the Tore underwater mountain range. The volcano formed atop an eastwest riff in the oceanic crust along the African plate, beginning during the Miocene epoch over 5 million years ago, continuing into the Pleistocene until about 700,000 years ago (Geldmacher et al. 2000). The most recent volcanic eruptions were on the west-central part of the island only 6,500 years ago, creating more cinder cones and lava flows (Carracedo et al. 1998).

Madeira has a mountain ridge that extends along the center of the island, reaching 1,862 m at its highest point (Pico Ruivo), while much lower (below 200 m) along its eastern extent. The primitive volcanic foci responsible for the central mountainous area consisted of the peaks Ruivo (1,862 m), Torres (1.851 m), Arieiro (1,818 m), Cidrão (1,802 m), Cedro (1,759 m), Casado (1,725 m), and Ferreiro (1,582 m). At the end of this eruptive phase, an island circled by reefs was formed; its marine vestiges are evident in a calcareous layer in the area of Lameiros, in São Vicente.

Also the seven major islands, one minor island, and several small islets of the Canary Islands were originally volcanic islands. The Teide volcano on Tenerife is the highest mountain in Spain and the third tallest volcano on Earth on a volcanic ocean island. All the islands except La Gomera have been active in the last million years; four of them (Lanzarote, Tenerife, La Palma, and El Hierro) have historical records of eruptions since European discovery.

The islands rise from Jurassic oceanic crust associated with the opening of the Atlantic. Underwater magnetism commenced during the Cretaceous and reached the ocean surface during the Miocene. The islands are considered as a distinct physiographic section of the Atlas Mountains Province, which in turn is part of the larger African Alpine System division (Del Arco Aquilar and Delgado 2018).

Zonal soil types have been defined based on different geologic substrates, numerous orographic landforms, and earlier geomorphological processes, as well as climatic factors and the development of the vegetation cover. The arctic zone shows arctic frost debris and tundra (gley) soils (Gelic Regosols, Gelic Histosols) and the boreal zone podzol soils (Podzols, Podzoluvisols), gley-podzols, and peat soils (Histosols). In the temperate zone main soils are brown earths (Cambisols) and para-brown earths (Luvisols) and also podzol soils on base-poor substrates. In the eastern forest steppe and steppe zones, gray forest soils (Luvisols) and chernozems are dominant. In the warm humid areas in the south of the Colchis, mainly yellow and red earths (Acrisols) occur. Typical soils of the Mediterranean sclerophyllous forests and scrub are cinnamon-colored, reddish-brown, and red soils (Calcic Cambisols, Chromic Luvisols). As climate and vegetation, soil types vary according to altitude (Kondracki and Schlüter 2003).

Macaronesian soils mainly developed from airborne volcanic deposits, especially basic ashes. They are rich in amorphous weathering products and volcanic glass and therefore finely coarse, with low bulk density, but nevertheless, they are usually quite slicky. The ion exchange capacity is rather high, which make these soils very fertile. The soils in Azorean islands are classified as Andosols (Schäfer 2002).

Climate is crucial for the zonal differentiation of vegetation types and soil formation. At the

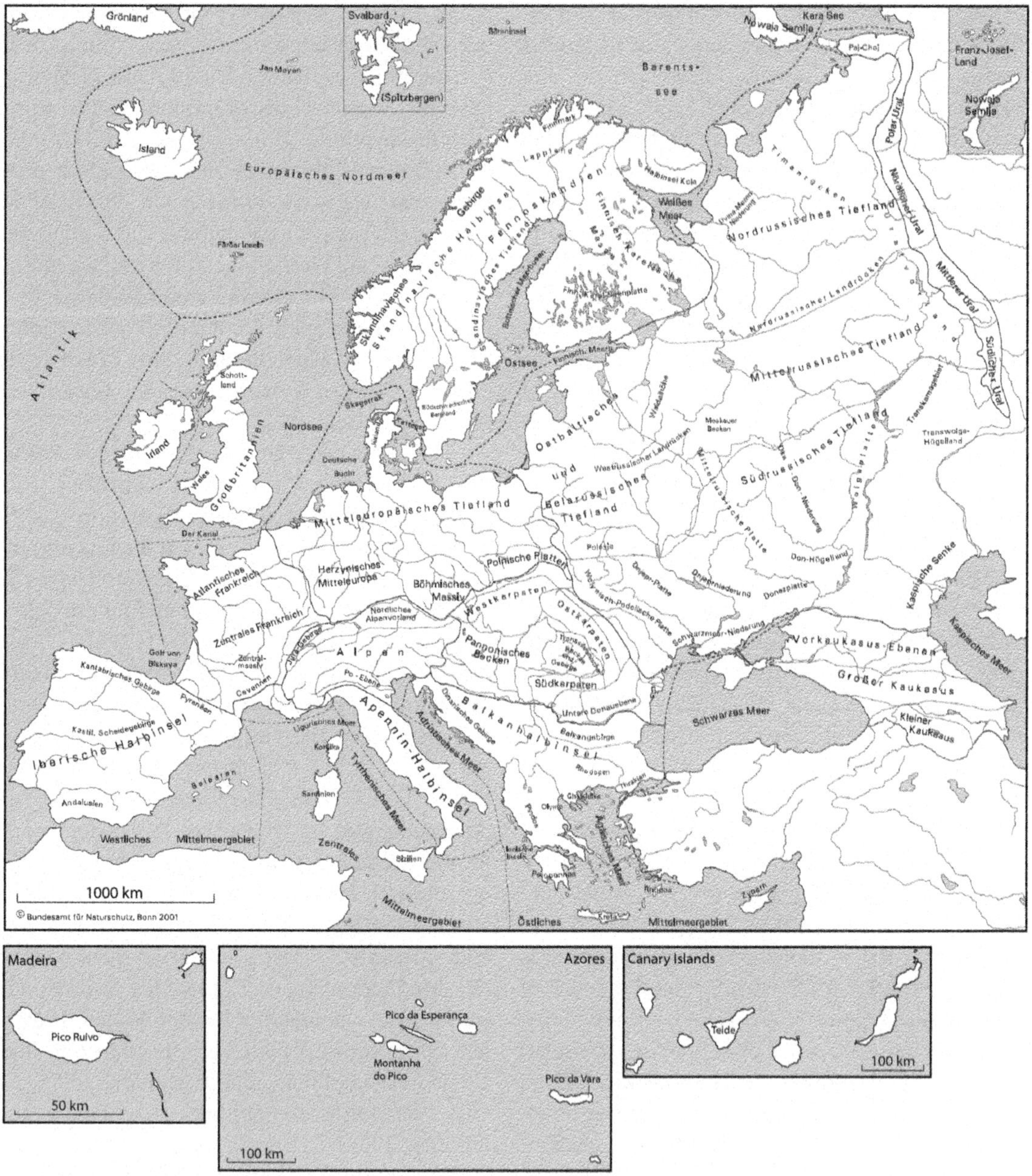

Fig. 3 Physiographical structure of Europe. (After Kondracki and Schlüter (2003), modified. Reproduced with permission from Bundesamt für Naturschutz/Federal Agency for Nature Conservation/2018)

same time, soil types, especially concerning moisture regime, nutrient status, and texture, have a great influence on the formation of regional vegetation types and mosaics and are helpful in the recognition of biotopes at landscape level.

Physiographical Structuring

Unless indicated otherwise, this chapter follows Kondracki and Schlüter (2003).

Natural landscapes are formed by numerous natural factors, altered by varying human influence. The orography and lithology of rocks are the most stable factors of landscapes. The most unstable factor is vegetation, which is also highly affected by human land use. As a result, the potential natural vegetation represents its natural composition, structure, and spatial distribution. The potential natural vegetation is the most prominent landscape-ecological feature and corresponds to the description of biotopes representing the integration of all natural landscape components.

For physiographical regionalization, the horizontal differentiation of climate, bedrock, and soil properties and the vertical sequence from lowland to hill country (colline), upland (montane), and high mountain (alpine and nival) regions are most important and have a great influence on the spatial differentiation of the natural vegetation. For our small-scale consideration of Europe, we only require the highest hierarchical spatial units, subdivided according to their overall physiographical character into super-region, region, and province (see Fig. 3).

Adriatisches Meer	Adriatic Sea
Ägäisches Meer	Aegean Sea
Alpen	Alps
Andalusien	Andalusia
Apennin-Halbinsel	Apennine Peninsula
Atlantik	Atlantic Ocean
Atlantisches Frankreich	Atlantic France
Balearen	Balearics
Balkangebirge	Balkan Mountains
Balkanhalbinsel	Balkan Peninsula
Bäreninsel	Bear Island
Barentssee	Barents Sea
Böhmisches Massiv	Bohemian Massif
Bottnischer Meerbusen	Gulf of Bothnia
Cevennen	Cévennes
Chalkidike	Chalkidiki
Der Kanal	(English) Channel
Deutsche Bucht	German Bight
Dinarisches Gebirge	Dinaric Alps
Dnjepr-Platte	Dnieper Upland
Dnjeprniederung	Dnieper Lowland
Donezplatte	Donbass
Don-Hügelland	Hills of River Don
Dvina-Mezen-Niederung	Dvina-Mezen-Valley
Färöer Inseln	Faeroe Islands
Fennoskandien	Fennoscandia
Finnischer Meerbusen	Gulf of Finland
Finnisch-Karelische Masse	Finnish-Karelian Massif
Finnisch-Karelische Seenplatte	Finnish-Karelian Lakeland
Finnmark	Finnmark
Franz Josef Land	Franz Joseph Land
Golf von Biskaya	Bay of Biscay
Grönland	Greenland
Großbritannien	Great Britain
Großer Kaukasus	Greater Caucasus
Halbinsel Kola	Kola Peninsula
Herzynisches Mitteleuropa	Hercynian Europe
Iberische Halbinsel	Iberian Peninsula
Irland	Ireland
Island	Iceland
Jura-Gebirge	Jura Mountains
Kantabrisches Gebirge	Cantabrian Mountains
Kara See	Kara Sea
Kaspische Senke	Caspian Depression
Kaspisches Meer	Caspian Sea
Kastil. Scheidegebirge	Central System (Sistema Central)
Kattegat	Kattegat
Kleiner Kaukasus	Lesser Caucasus
Korsika	Corsica
Kreta	Crete
Lappland	Lapland
Ligurisches Meer	Ligurian Sea
Mitteleuropäisches Tiefland	Central European Lowlands
Mittelrussische Platte	Central Russian Upland
Mittelrussisches Tiefland	Central Russian Lowlands
Mittlerer Ural	Central Ural
Moskauer Becken	Moscow Basin
Nördlicher Ural	Northern Ural
Nördliches Alpenvorland	Northern Alpine Foothills
Nordrussischer Landrücken	Northern Ridge, Northern Uvaly
Nordrussisches Tiefland	North Russian Plain
Nordsee	North Sea
Nowaja Semlja	Novaya Zemlya
Oka-Don-Niederung	Oka-Don Lowlands
Olymp	Mount Olympus
Ostbaltisches und Belarussisches Tiefland	East Baltic and Belarusian Plain

(continued)

Ostkarpaten	Eastern Carpathians
Östliches Mittelmeergebiet	Eastern Mediterranean
Ostsee	Baltic Sea
Paj-Choj	Pay-Khoy Ridge
Pannonisches Becken	Pannonian Basin
Pecora-Niederung	Pecora Valley
Peloponnes	Peloponnese
Pindos	Pindos
Po-Ebene	Po Valley
Polesje	Polesie
Polnische Platten	Poland Uplands
Pyrenäen	Pyrenees
Sardinien	Sardinia
Schwarzes Meer	Black Sea
Schwarzmeer-Niederung	Black Sea Lowland
Sizilien	Sicily
Skandinavische Halbinsel	Scandinavian Peninsula
Skandinavisches Gebirge	Scandinavian Mountains
Skandinavisches Tiefland	Scandinavian Lowland
Spitzbergen	Svalbard
Südkarpaten	Southern Carpathians
Südlicher Ural	Southern Ural
Südrussisches Tiefland	South Russian Lowlands
Thrakien	Thrace
Timanrücken	Timan Ridge
Transkamagebiet	Trans-Kama region
Transwolga-Hügelland	Transvolga
Tyrrhenisches Meer	Tyrrhenian Sea
Untere Donauebene	Lower Danubian Plain
Uralvorland	Ural foothills
Vorkaukasus-Ebenen	North Caucasus region
Waldaihöhe	Valdai Hills
Weißes Meer	White Sea
Westkarpaten	Western Carpathians
Westliches Mittelmeergebiet	Western Mediterranean
Westrussischer Landrücken	Belarusian Ridge
Wolgaplatte	Volga Upland
Wolynisch-Podolische Platte	Podolian Upland
Zentrales Frankreich	Central France
Zentrales Mittelmeergebiet	Central Mediterranean
Zentralmassiv	Massif Central
Zypern	Cyprus

According to a convention of the International Federation for Documentation (FID 1971), Europe is physiographically divided into nine super-regions. A useful overview can be obtained by first subdividing the continent into four parts following the cardinal directions. This results in four subcontinents to which the nine super-regions can be related. Macaronesia will be added as the 10th super-region.

Northern Europe

1. Fennoscandia, Iceland, Arctic islands

Western and Central Europe

2. British Isles and France
3. Northern Central Europe
4. Alps
5. Carpathian Mountains

Southern Europe

6. Mediterranean Southern Europe

Eastern Europe

7. Caucasus and Crimea
8. East European Lowland
9. Ural Mountains

Macaronesia

The Super-Regions of Europe (Following Kondracki and Schlüter 2003)

Northern Europe (approximately 1,500,000 km^2) is separated from the mainland of the continent by several waters: the North Sea, Skagerrak, Kattegat, the Baltic Sea with the Gulf of Finland, Lake Ladoga and Lake Onega, the White Sea, and the Barents Sea. It can be divided into two large parts of different geological origin, the Precambrian shield of Fennoscandia and the early Paleozoic (Caledonian) folding of the Scandinavian Mountains. Another part of Northern Europe are the Arctic islands Novaya Zemlya, Franz Josef Land, Svalbard, Bear Island, Jan Mayen, and Iceland, although the latter is of oceanic-volcanic and not continental origin.

Quaternary glaciation marked Northern Europe's surface significantly with glacial erosion leaving basins filled by lakes or mires and deeply cut fjords in the landscape on the western side of the Scandinavian Peninsula. Mountain ranges in the southern part stand out with altitudes exceeding 2,000 m above sea level (Galdhøpiggen 2,469 m) and extensive glaciers (Jostedalsbreen, approximately 1,000 km^2).

The climate in Northern Europe is influenced by the warm Atlantic Gulf Stream with maritime air causing moist, less severe winters, cool summers, and precipitation distributed over the whole year. In the western and southern part of the peninsula, this oceanic influence is restricted by the Scandinavian Mountains which stretch across almost 2,000 km from the southwest to the northeast. Thus the northeastern parts have a more continental climate with very cold winters, as in Finland and Karelia (mean January temperatures of −8 °C to −16 °C).

Four phytogeographical zones can be distinguished in Northern Europe, from north to south: arctic polar deserts and tundras, boreal birch and coniferous forests (taiga), hemiboreal mixed broadleaved-coniferous forests, and nemoral broadleaved forests in the south and southwest (Kondracki and Schlüter 2003).

Northern Europe is subdivided as follows:

1 *Fennoscandia, Iceland, Arctic islands*
11 Iceland
12 Jan Mayen Island
13 Faeroe Islands
14 Scandinavian Peninsula with its offshore islands
15 The Finnish-Karelian Shield
16 The Kola Peninsula
17 Svalbard and Bear Island
18 Franz Joseph Land
19 Novaya Zemlya

Western and Central Europe (approximately 2,250,000 km^2) covers the western part of the Eurasian mainland with the British Isles and shows a complicated geological composition which has been broken up by manifold faults, uplifts, and sinkings, in particular during the periods of mountain formation which include the Caledonian, Hercynian, and Alpine foldings. In the course of these processes, during a sinking between Northern and Western Europe, the North and Baltic seas area and the Central European lowlands developed. Today the lowlands are covered by Tertiary and Quaternary sediments.

Several uplands formed as a result of the foldings. On the British Isles, these are the Scottish Highlands (1,343 m) and the uplands in Cornwall and Wales (1,085 m) that extend as far as the south and southeast of Ireland. On the European mainland, mountainous areas extend from France across Germany to Poland: one arch beginning in Brittany, France, extending to the Massif Central (Puy de Sancy 1,886 m), and a second area extending from there to the northeast: the Ardennes (694 m), the Rhenish Slate Mountains (841 m), the Vosges (1,424 m), the Black Forest (1,493 m), and the Harz (1,142 m). The Bohemian Massif with its prominent ridges, namely, the Bavarian and the Bohemian Forests (1,457 m), the Thuringian Forest (982 m), the Erzgebirge (1,244 m), and Sudeten mountains with the Giant Mountains (1,603 m), is a special mountainous area in the eastern part.

The mountain foldings were often accompanied by the formation of intermontane basins which often also represent special climatic and phytogeographical conditions as in continentally influenced dry areas. Examples for such areas are the Thuringian Triassic basin (annual precipitation <500 mm) with the adjacent Magdeburger Börde to the east (chernozems) or the Prague Basin.

The Alps and the Carpathian mountains form the third orographic element of Central Europe. These are the high mountain ranges including their respective basins both within and around the mountains. The Western and Central Alps are higher and more "compact" with several peaks over 4,000 m (Mont Blanc 4,807 m) than the Eastern Alps, which feature no peaks higher than 4,000 m (Großglockner 3,797 m). Pleistocene glaciations shaped the alpine relief significantly and also led to the formation of large morainal ridges and lakes (Lake Constance, Lake Geneva, Lake Garda, Lake Maggiore, etc.). The second high mountain range in the northeast is the Carpathian Mountains, which exceed 2,500 m only in their western and southeastern parts (High Tatras 2,655 m, Fagarash Mountains 2,543 m). The Pannonian Basin is more than half surrounded by the Carpathian Arch. In the southeast, the rivers Sava and Lower Danube define the border of Central Europe, where they divide the Pannonian and Carpathian regions from the Balkan Peninsula.

Pleistocene sediments are the main components of Central European lowlands consisting of glaciogenic terminal and ground moraines as well as glaciofluviatile and fluviatile sediments, with altitudes which rarely exceed 200 m and only reach 300 m in exceptional cases. In the east Central Europe's border is defined climatically and biogeographically, in the northwest from Eastern Poland through Western Ukraine to the southeast in Eastern Romania.

Central Europe lies between 43° and 58°N and belongs to the relatively warm temperate zone with deciduous broadleaved forests. There is a notable climatic differentiation although climatic zonation is rather indistinct. The strongly differentiated orography and the influence of the seas in the north and south have a strong impact on climatic characteristics. There are vegetation zones affected by increasing dryness of climate up to in part the presence of forest steppe as in the Pannonian Basin, the Romanian Danube lowlands and the eastern foothills of the Carpathian Mountains. Another dominant climatic zone is moderately warm and moist suboceanic with decreasing precipitation and larger ranges in seasonal temperature with increasing continentality toward the east. These are the areas north of the Alps and Carpathian mountains. The mountain ranges are dominated by altitudinal belts with both climatic conditions and vegetation types changing according to altitude. There is a considerable rise in precipitation and drop in temperature on a yearly average as altitude increases.

The Western and Central European super-region is subdivided into the following physiographical regions and provinces:

2 *British Isles and France*
21 Ireland
22 Great Britain and neighboring islands
23 a Atlantic France
23 b Central and Southern France
3 *Northern Central Europe*
31 Central European lowlands
32 Hercynian Central Europe (low mountains and cuesta landscape)
33 The Bohemian Massif and its surrounding mountain ranges
34 The Polish uplands
4 *Alpine Countries*
41 Jura mountains
42 Northern foothills of the Alps
43 Alps
44 North Italian lowlands
5 *Carpathian Countries*
51 Western Carpathian Mountains and outer foothills
52 Eastern Carpathian Mountains and outer foothills
53 Southern Carpathian Mountains and outer foothills
54 Transylvanian basins and mountains
55 Pannonian Basin
56 Lower Danube plain

Southern Europe (approximately 1,250,000 km^2) is not a contiguous mainland area; it is composed of the three large Mediterranean peninsulas with their neighboring islands. The formation of the peninsulas is closely connected to the orogenesis of the relatively recently formed Alpine mountain chains. The rather compact Iberian Peninsula is delimited in the north by the Pyrenees (Pico de Aneto 3,404 m) and the Cantabrian Mountains and in the southeast by the Baetic Cordillera with the Sierra Nevada (Mulhacen 3,478 m) which continues on to the Balearic Islands. The islands of Corsica and Sardinia are remnants of the sunken Tyrrhenian Massif. The Apennine mountain chain (Corno Grande 2,914 m) extends along the Apennine Peninsula and forms the peninsula lengthwise, continuing on to Sicily. The Balkan Peninsula is formed in the center by the older Thracian-Macedonian Massif (Musala in the Rila Mountains 2,925 m, Olympus 2,918 m, the

Pindos with Smolikas 2,637 m, Nkiona 2,510 m) which is surrounded by the Dinaric Mountains (Prokletije 2,693 m) and the Balkan Mountains (Botev 2,376 m). Its southern part has been tectonically broken up into peninsulas such as the Peloponnese and Chalcidice as well as the Aegean Islands (Kondracki and Schlüter 2003).

A peculiarity of Southern Europe is its seismic instability with the active volcanoes Mount Etna (3,340 m) on Sicily, the Aeolian island Stromboli with Mount Stromboli (926 m), and Mount Vesuvius (1,277 m), all located on the Apennine Peninsula. These are the only volcanoes still active on, resp., near the European mainland.

The characteristics of Southern Europe climate, belonging to the Mediterranean climate type, are warm and dry summers as well as moist and almost frost-free winters. Primarily this applies to lowlands and coastal areas of all three peninsulas and the Mediterranean Sea with its islands. Because of the dominating mountain ranges on the peninsulas, in inland areas, continentality tends to increase, particularly on the Iberian and Balkan Peninsulas.

Due to the predominant climate and the closely related soil types, vegetation, and drainage regime, the Mediterranean area differs quite strongly on the whole from Western and Central Europe. Mediterranean Southern Europe is subdivided into the following regions:

6 *Mediterranean Southern Europe*
61 Iberian Peninsula with Balearic Islands
62 Central Mediterranean area
63 Apennine Peninsula
64 Balkan Peninsula and neighboring islands
65 The Balkan Mountains

Eastern Europe (approximately 5,000,000 km^2) with its vast, topographically monotonous landscape differs distinctly from the rest of Europe. This large plain is characterized by the level position of sedimentary rocks on the Precambrian basement which is exposed in Fennoscandia (the Baltic Shield) and in the Ukrainian Shield. Extending eastwards to the Urals are basins covered by Paleozoic and Mesozoic strata. The Devonian Timan Ridge in the northeast, the Kursk Heights with crystalline bedrock at the upper Don, and the Hercynian folding zone in the Dnieper-Donets area represent particularities in the geological structuring of Eastern Europe. The east European lowlands end at the Ural Mountains range chain, while the southern boundary is formed by the northern Black Sea coast, the Crimean Mountains and the Great Caucasus.

The Crimean Mountains (Roman-Kosch 1,545 m) are asymmetrically formed with a steep slope to the south toward the Black Sea. As a result, the coastline is protected from cold air in winter, and a "Riviera" has developed with an almost Mediterranean climate. Further to the east, the immense chain of the Great Caucasus stretches to the Caspian Sea, with several extinct volcanoes higher than 5,000 m (Elbrus 5,633 m). In the east, the Ural Mountains extend in a north-south direction for approximately 2,500 km. The northern part has the highest summits (Narodnaja 1,894 m); the middle part is slightly lower before rising again up to 1,640 m in the southern part.

A number of ridges and plateaux are found in the south of the east European lowlands. These include the Volyn-Podilsk Upland (Kamula 474 m), the Dnieper, Donets, Central Russian, and the Volgian uplands, as well as the Yergeni Hills in the south. Lowland plains are the dominant landscape feature which extend between the ridges and plateaus and further south. These include the Trans-Dnieper, the Oka, Don, as well as the Black Sea and the Caspian lowland in the south and also extensive lowlands, e.g., the Polesje and low uplands such as the West Russian Ridge, the Valdai Hills, the Central Russian Uplands, and the North Russian Ridge further north.

As it is in the Central European lowlands, the topography of the northern part of Eastern Europe is of glacial origin. Glaciers advanced the furthest during the so-called Dnieper Glaciation in the Middle Pleistocene and covered the landscape as far as the Polesje and the Dnieper lowland to Dnipropetrovsk and the Oka-Don lowland to Kalach-na-Donu. The undulating lake landscape in a zone connecting Vilnius, Vitsyebsk, and the Valdai Hills (343 m) and extending to the northeast to Archangelsk on the White Sea was created during the last so-called Valdai glacial period. Periglacial loess deposits are found in the uplands of the southern part of Eastern Europe, while the north Caspian lowland is covered by saline sand and clay sediments dating from the Pleistocene transgressions over the Caspian Sea.

The rather flat, vast land surface of Eastern Europe stretches out across approximately 25° of latitude. This area is also characterized by pronounced geographic zonation. Continentality increases toward the east and causes arid regions in the south and cold regions in the north. The southwest to northeast trend of zonal boundaries results in a narrowing of the mixed broadleaved forest zone toward the east and its complete disappearance east of the Urals.

For the physiographic subdivision of Eastern Europe, climatic, edaphic, phytogeographical, and geomorphological criteria must be taken into account. The classification lists these subdivisions as follows (Kondracki and Schlüter 2003):

7 *Caucasus and Crimea*
71 Crimean Peninsula
72 Pre-Caucasus plains
73 Great Caucasus
74 Colchidian (Rioni) lowlands (belonging to Asia Minor)
75 Kura lowland (belonging to Asia Minor)
76 Lesser Caucasus (belonging to Asia Minor)

8 *East European Lowland*
81–82 North Russian lowland
83 Central Russian lowland
84 East Baltic and Belarusian lowland
85–86 South Russian lowland

9 *Ural Region*
91 Paj-Choj Ridge and Vaygach
92 Polar Urals
93 Northern Urals
94 Central Urals
95 Southern Urals

To add Macaronesia as "part of Europe," the following subdivisions are proposed:

10 *Macaronesia*
101 Azores
102 Madeira Archipelago
103 Canary Islands

Climatic Classification of Europe (Following Walter et al. 1975)

Climatic conditions in Europe are defined by geographical location, landmass distribution, and topography. Five main thermal zones can be distinguished from north to south: arctic, boreal, temperate, submeridional, and meridional. These relate closely to the southward progression of zonal vegetation from arctic tundra through boreal coniferous forests (taiga) to temperate mesophytic and thermophilous broadleaved deciduous woods as well as the steppes and deserts of Eastern Europe and finally to Mediterranean evergreen sclerophyllous forests and scrub.

Geographical latitude and regional differentiation of the surface relief have an effect on the climate mainly in the form of the solar radiation balance, but the climate is also influenced by atmospheric circulation. Northern Europe is warmer than would be expected regarding its latitude. That is because of the westerly circulation and the Gulf Stream; nevertheless very cold arctic air masses can at times reach the area. In the temperate zone, this westerly circulation moderates the temperature leading to relatively mild winters and comparably cool summers. Sometimes, however, dry continental or cold arctic air masses advance to Northern Europe and lead to a great instability in weather patterns and make the course of weather vary considerably over the year. The atmospheric circulation is also influenced continuously by energy input from the sun that varies according to multiphase rhythms (11-year, 35-year and longer cycles between cold-moist and warm-dry periods). In the Mediterranean region, in summer typically warm and dry weather are predominant due to the anticyclone acting as a high-pressure area, while in winter depressions entering from the west bring substantial precipitation. Sometimes this circulation pattern can also reverse (Kondracki and Bohn 2003).

Oceanic air masses warmed by the Gulf Stream mainly affect Western and Northern Europe (despite its gradual weakening), and a warm subtropical North Atlantic Ocean current influences temperatures all the way up to the Kola Peninsula. This is a fundamental aspect of the European climate. Even in the far north, the Gulf Stream moderates the long arctic winters, so that the port of Murmansk north of the Arctic Circle can remain ice-free during the severest of winters.

In Eastern Europe in winter, the Asiatic anticyclone produces a flow of cold, dry air masses causing an increase in the typical characteristics of continental climate from Western to Eastern Europe. This climatic particularity includes an increase in the annual range of average monthly temperatures and a shift toward a summer precipitation maximum.

In the oceanic climates on the western edge of Europe, by contrast, the amount of precipitation that falls per month is almost the same in every month of the year, and the annual amplitude between mean monthly temperatures is small. This implies that winters tend to be mild and frost-free, and summers are relatively cool.

Precipitation is highest in the western part of the temperate climate zone, decreasing from west to east with increasing continentality. Accordingly, mean annual precipitation varies from West to East, Valentia (SW Ireland) with 1,416 mm (with a winter maximum), London with 612 mm, Warsaw 550 mm, and Voronezh (Russia) with 521 mm (the last two places have a summer maximum). In the arctic north, it is even less, in Murmansk (Russia), for example, with 477 mm. In the Mediterranean climate region precipitation differs greatly. In Palma on Mallorca, it is low (498 mm) as it is in Athens (407 mm), while in areas affected by the westerly circulation, it is much higher, with a marked winter maximum, for instance, Genoa receives 1,258 mm, while Rijeka receives 1,593 mm.

Annual precipitation also increases with increasing altitude. In some mountain ranges, it can exceed 4,000 mm (e.g., the Dinaric and Scandinavian mountains, Wales), while in the Alps it is more than 2,000 mm. Mean annual temperature, by contrast, decreases continuously with increasing altitude.

In the eastern part of Europe, the north-south climatic zonation shows most notably, as it represents a contiguous mainland block that is mostly flat and not broken up by mountain ranges or large bodies of water. Temperature and moisture balance considered, the following vegetation-related main zones can be distinguished (Kondracki and Bohn 2003):

Arctic tundra zone
Boreal birch and coniferous forest zone (taiga)
Temperate coniferous-deciduous forest and deciduous forest zone
Submeridional-continental, periodically dry steppe zone
Submeridional arid, continuously dry desert zone

Between these five main zones, the following transitional sub-zones or subregions can be distinguished: subarctic forest tundra, hemiboreal deciduous broadleaved-coniferous forests, subcontinental forest steppes, and semiarid desert steppes. The western part of Europe shows climatic zones that are not as pronounced because of the compensating oceanic influence and the horizontal and vertical structure of the landscape.

The horizontal and vertical differentiation of the climatic factors and the vegetation types that depend on them are crucial in the physiographical classification of the continent.

Ecological Classification of the Climates of Europe

The mapping units and formations of the vegetation map of Europe are related to various climatic types based on the World Atlas of Climate Diagrams (Walter and Lieth 1967). The climatic data (annual precipitation, annual and monthly mean temperatures) for the mapping units were primarily taken from the climate diagrams published in that work and in part also from other publications.

A simplified world map, "ecological climate classification of the continents," was published in 1975 by Walter, Harnickell, and Müller-Dombois (Climate diagram maps of the individual continents and the ecological climatic regions of the earth). The climate zones for Europe are displayed in Fig. 4.

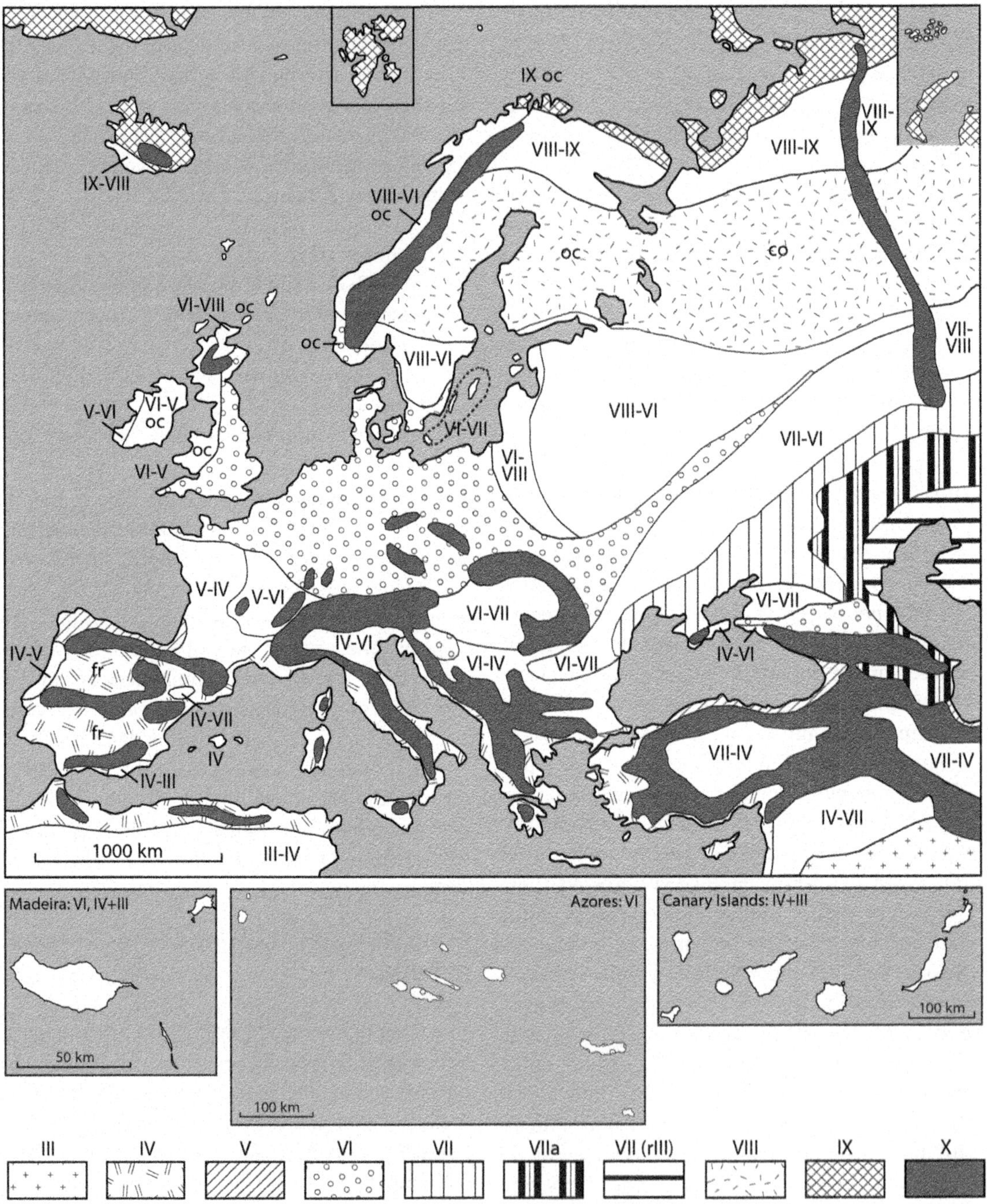

Fig. 4 Climatic classification of Europe. III: Subtropical arid zone of the deserts; IV: Typical Mediterranean climate; V: Warm temperate climate; VI: Typical temperate climate; VII: Semiarid steppe climate; VIIa: Arid semi-desert climate; VII(rIII): Desert climate with cold winter; VIII: Cold-temperate or boreal climate zone; IX: Arctic climatic zone; X: Mountain climates. VI–VII refers to a transition between VI and VIIV (IV) corresponds to a warm temperate climate (V) but with prevalent winter rains (as with IV). VIII–IX: Subarctic zone (the delimitation of zones IX and VIII is based on the duration of the vegetationperiod with daily averages above 10 °C. Its duration at the northern boundary of zone VIII is about 30 days, while at the southern limit it is about 120 days). VIII–VI: Hemiboreal zone VI–VII: Moderately arid temperate, subcontinental forest-steppe zoneVI–IV: Temperate sub-Mediterranean zone V–VI: Submediterranean zone IV–III: Thermomediterranean arid zone. Additional symbols: oc = oceanic; co = continental; fr = frequent frost. (Modified from Walter et al. 1975)

Explanation of the Climate Map (According to Walter et al. 1975)

The map shows the ranges of the climate types I–IX, i.e., the regional climates of large areas with similar topography (mainly lowland, hill country, and lower mountains). The mountain climates have been presented as a single unit (X) without subdivision into altitudinal belts or assignment to the major climatic zones. Transitional zones between the individual climate types are indicated by combinations of the relevant Roman numerals. Climatic gradients (temperature and precipitation) also exist within the ranges of the individual climate types from west to east and from north to south. Variations from the norm are marked with additional characters from the alphabet (a, oc, co, fr, r).

The Macaronesian climate seems to be a mosaic of oceanic temperate climate (VI), warm temperate climate (V), typical Mediterranean (IV), and subtropical arid (III). Del Arco Aguilar and Delgado (2018) differentiate bioclimatic belts in the Canary Islands: intra-Mediterranean, thermo-Mediterranean, meso-Mediterranean, supra-Mediterranean, and oro-Mediterranean. Important is the presence or absence of trade wind clouds; therefore in all belts, ombrotypes from hyperarid, arid, semiarid, dry, subhumid, to humid can exist.

The Azorean climate is temperate oceanic with a mean annual temperature of 17.5 °C and mean annual precipitation between 1,000 and 1,600 mm at sea level. The climate is mainly regulated by a branch of the Gulf Stream and by a high-pressure zone called the "Azores anticyclone." The influence of the anticyclone is at its highest in summer. The average snow line is located at 1,200 m, but may in some exceptions reach as low as 600 m. Rainfall peaks during the time from October to January and is at its lowest in July. Throughout the Azores, rainfall decreases from W to E (Harris et al. 1962). Cloudiness is more common than in any other Macaronesian region, and fog on the islands is also a rather common phenomenon. The influence of NE trade winds is most important in the easternmost island, whereas winds from N, NW, and SW prevail in the central and western groups (Schäfer 2002; Cropper and Hanna 2013).

The climate on Madeira Islands ranges from temperate oceanic in the North to subtropical arid in the South. During summer, the NE trade winds influence the climate on the Canary Islands. Average temperatures are between 15.5 °C (minimum) and 21.9 °C (maximum); the average precipitation is 642 mm.

On the Canary Islands, according to the position of the islands with respect to the northeast trade winds, the climate can be mild and wet or very dry. As a consequence, the individual islands in the Canary archipelagos tend to have distinct microclimates. Those islands lying to the west of the archipelago, such as El Hierro, La Palma, and La Gomera, have a climate which is influenced by the moist of the Gulf Stream. Basically the climate is Mediterranean but can be more humid or dry depending on the position of the islands. Generally, the Canaries have hot, dry summers and warm, humid winters, but locally there are often considerable deviations from this basic pattern (Del Arco Aquilar and Delgado 2018).

The northeast trade winds bring in moisture from the sea, which, when forced to rise by the mountain barriers of the western islands, is cooled and forms a zone of precipitation at about 800 to 1,500 m. This causes a more or less persistent cloud layer at this level on the north side of all western islands and has a great influence on the natural vegetation. The southern sectors of these islands are in a rain shadow and receive much less precipitation. They are generally without a tense-forest zone at mid-altitude level and are much more xerophytic in nature. The climate is more humid and the vegetation more luxuriant. The further west one goes in the archipelago due to the increasingly oceanic position of the western islands and the greater strength of the trade winds (Wirthmann 1970; Cropper and Hanna 2013).

The extreme dryness of the eastern islands and the south of Gran Canaria is partially due to the hot dry Saharian winds, the Levante, which sometimes reach the eastern Canaries blowing for up a week at a time with a strong desiccating effect on the vegetation. The eastern islands are too low to intercept the trade wind moisture except for their highest points (Del Arco Aquilar and Delgado 2018).

Phytogeographical Division of Europe

The division of the Earth into floristic kingdoms, regions, and provinces intends to classify areas with a floristic inventory as homogeneous as possible. The division of Europe presented here was also established for this purpose (Meusel and Jäger 1992). It has already turned out to be useful for characterizing plant species ranges on many occasions.

Unlike vegetation maps, a floristic division is based on the limits of plant distribution ranges and floristic gradients in particular of higher plants. Cryptogams are not considered as their ranges are not yet known well enough. The floristic gradient describes the density of different range limits observed in a single direction within a given distance. The position and distinctiveness of a phytogeographical boundary are defined on the basis of the floristic gradient.

It can be expected that floristic division corresponds largely to the boundaries of principal vegetation types on a continental scale. However, on a global scale, isolation due to the changing distribution of sea and land may cause major floristic differences between areas that show similar types of vegetation due to the effects of convergent evolution (Jäger 2003).

It is not yet possible to establish an objective floristic division that takes into account the ranges of all species due to a lack of data, even for a well-studied area like Europe.

The floristic divisions developed by Meusel and Jäger (1992) are mainly based on regularly occurring distribution limits. In addition to comparing hundreds of distribution maps, all classification proposals of earlier authors were considered (including those proposed in floras).

Methodological problems turn up with all phytogeographical divisions. Phytogeographers find floristic limits and floristic influences from various directions in their narrow working area. It is therefore not surprising that almost everyone speaks of "an interesting transition area." However, depending on the floristic contrast, the floristic boundaries vary in importance and also differ in clarity depending on the floristic gradient. Clear floristic boundaries are quite obvious with steep gradients in physicogeographical factors (Jäger 2003).

Looking at differences in floristic contrast, a hierarchy of floristic areas soon becomes evident. When demarcating the phytochores (floristic areas) of higher rank (floristic kingdoms, floristic regions), the systematic rank of the characteristic taxa is especially emphasized. Thus, for defining floristic kingdoms, several endemic families should be present and for floristic regions, several endemic genera, while floristic provinces must be characterized at least by the presence of a few endemic species.

Floristic division refers to lowland and colline belts as well as to a uniform type of sequence of mountain altitudinal belts in the area considered. However, this is not possible in expansive highland areas, for example, in Central Asia. The Alps, Carpathians, and Caucasus were ranked as subregions in their own right (extending over two zones) as they have many of their own species. The subdivision of the Alps varies depending on the different altitude belts. The montane belt is more homogeneous in composition than the lower ones (Hübl and Niklfeld 1973).

Since plant distribution is influenced by climatic changes, all floristic area limits are dynamic. With winter temperature and precipitation rising at more northern latitudes, a major increase in oceanicity is notable in this region. With an increase in average January temperatures by around 3 °C and summer temperatures by about 1 °C, for instance, the British Isles would be excluded completely from the potential range of *Tilia cordata*, while a rise in the July average by about 3 °C and the January average by about 1 °C would enlarge the potential range of this tree into Ireland. Former changes in climate as the medieval warmings (500–600 and 1,000–1,200 A.D.) and the "Little Ice Age" (1,550–1,700; 1,820–1,860 A.D.), supposedly accompanied by comparable temperature changes, may have led to similar effects (Crawford 2000). The amount of precipitation as well as the changing seasonality of rainfall controls plant distribution. The associated variable leaching and wetting of the soils and also the impact on the permafrost range crucially influence the range limits and the distribution of the natural vegetation (Jäger 2003).

Floristic Zones and Regional Subdivision (Fig. 5)

The arctic zone is made up of polar deserts and arctic tundras of Europe. The southern limit of this zone corresponds in Fennoscandia and Russia with the arctic tree line. This zone is specified by circumpolar species such as *Poa arctica*, *Luzula confusa*, *Cerastium regelii*, and *Draba corymbosa*.

The boreal zone commonly includes the taiga (i.e., northern coniferous forest) areas but in highly oceanic areas, e.g., in Iceland, on the Faeroe Islands, and in Western Norway, also birch forest areas and forest-free areas in which boreal herbaceous perennials and dwarf shrubs occur. Typical species for this zone are *Calypso bulbosa*, *Listera cordata*, *Rubus arcticus*, *Empetrum hermaphroditum*, *Linnaea borealis*, *Ledum palustre*, and *Cornus suecica*.

In the temperate zone, the formation of deciduous broadleaved forest extends from the Atlantic to the Urals, although in the north, taiga elements are interwoven, and in the south, islands of steppe vegetation occur. East of the Urals, small-leaved steppic forests with birch, aspen, and pine are set in the temperate zone, a view supported by the ranges of steppe forest plants which are distributed mainly temperately also in Central and Eastern Europe (*Dracocephalum ruyschiana*, *Crepis praemorsa*, *Hypochaeris maculata*, *Tragopogon pratensis* s. l.). The southern limit of the temperate zone is marked in entire Eurasia by, e.g., *Calamagrostis arundinacea*, *Carex vaginata*, *C. limosa*, *C. lasiocarpa*, and *Calla palustris*.

The oceanicity gradient is very steep in southern zones. In Europe, large parts of the suboceanic areas of the submeridional zone are inhabited by Mediterranean sclerophyllous vegetation. In parts of the central sub-Mediterranean province group (the Apennine and Balkan peninsulas) and in the eastern sub-Mediterranean province group (northern Anatolia and the Crimea to the Caucasus), species-rich deciduous woods predominate because of cold winter and moist summer conditions. In continental, drier areas of the zone, they are gradually replaced by forest steppe complexes and in the central and eastern Pontic zone by steppes. The southern boundary of the submeridional zone is marked in these different regions by the ranges of *Artemisia absinthium*, *Eleocharis acicularis*, or *Carex riparia*, for example.

The southern boundary of the meridional zone corresponds with the boundary of the Holarctic floristic kingdom. The range limits of many taxa of higher rank (Betulaceae, Fagaceae, Salicaceae, Platanaceae, Ranunculaceae, Paeoniaceae, Brassicaceae, Rosaceae, Aceraceae, Primulaceae) determine this boundary. Among the species ranges, there are only a few that can overcome the large span of oceanicity/humidity along this limit. But these species connect the very different vegetation areas at the southern edge of the Holarctic. These consist mainly of elements of azonal vegetation, i.e., aquatic and semiaquatic plants and rock plants that are less influenced by precipitation fluctuation, especially if they are salt-tolerant, as well as plants with synanthropic ranges (*Plantago major*, the genus *Carduus*, *Descurainia sophia*). In west Eurasia, *Bromus tectorum*, *Poa bulbosa*, and the genera *Colchicum* and *Helianthemum* mark the southern boundary of the meridional zone. The characteristic elements of this zone usually disappear to the north only in the submeridional zone (*Cistus*, *Daphne oleoides* group, *Arbutus*, etc.). The Macaronesian archipelagoes will be added to this zone.

Floristic Regions of Europe (Meusel and Jäger 1992)

Europe participates in five floristic regions. The Circumarctic and the Circumboreal floristic regions are circumpolar because of recent possibilities of plant dispersal to North America. The boundaries of these floristic regions correspond to the floristic zone boundaries with the same name.

The entire Middle European region with the Atlantic, sub-Atlantic, Central European, and Sarmatic floristic provinces belongs to temperate Europe. Together with the western, central, and eastern sub-Mediterranean provinces, it represents the European broadleaved deciduous forest area. In the Middle European region, a mix of relatively species-poor deciduous forest flora with numerous elements of Mediterranean origin is typical. The excellent growth and naturalization of many non-native laurophyllous woody plants

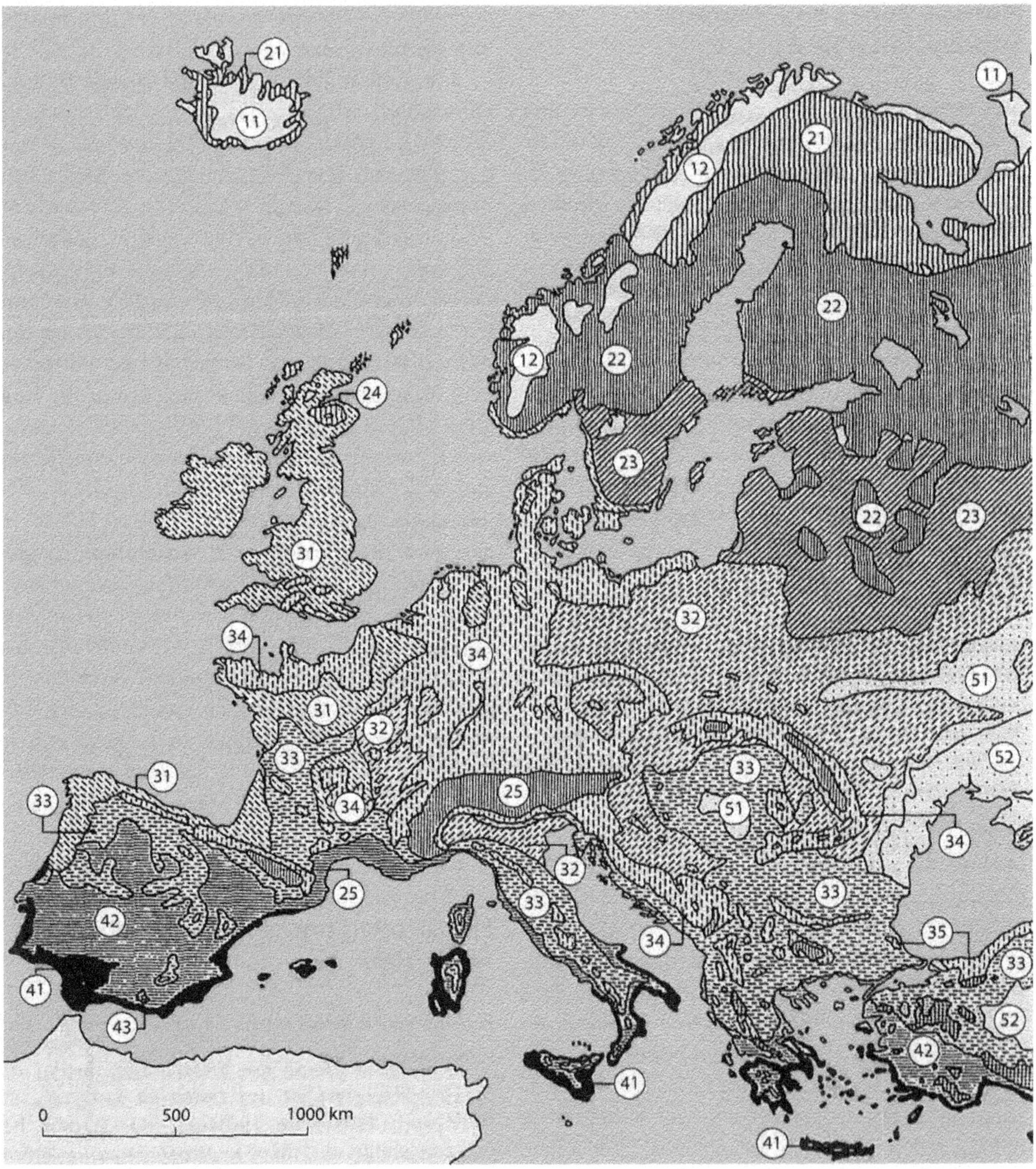

Fig. 5 Actual natural vegetation in a simplified overview. After Pfadenhauer and Klötzli (1994). 1. Polar and subpolar zone (11 = arctic tundra, 12 = alpine tundra). 2. Cold-temperate (boreal) zone (21 = deciduous boreal forests of *Betula pubescens* var. *pumila*; 22 = evergreen boreal conifer forests; 23 = hemiboreal mixed forests of broadleaved wood, spruces, and pines; 24 = boreoatlantic dwarfshrub heathland; 25 = montane and subalpine conifer forests and shrubs). 3. Humid cool temperate (nemoral) zone (31 = Western and Northwestern European oceanic mixed oak forests; 32 = central and Eastern European, subcontinental mixed oak forests; 33 = sub-Mediterranean mixed oak forests; 34 = western, central, and Southern European copper beech and copper beech-fir forests; 35 = Euxinian oriental beech forests). 4. Winter damp, warm moderate (subtropical) zone (41 = thermo-Mediterranean sclerophyll forests and scrubs; 42 = meso-Mediterranean holly oak and kermes oak forests). 5. Arid cool temperate (nemoral) zone (51 = Eastern European forest and meadow steppes; 52 = Eastern European high and short grass steppes). (Reproduced with permission from Springer Nature/2018)

(e.g., *Rhododendron ponticum*) in the Atlantic province is an indication that laurophyllous woody plants and oceanic conifers might dominate in this area, had these taxa not died out due to the late Tertiary-Quaternary cooling and aridization (Jäger 2003).

The south Siberian-Pontic-Pannonian region is part of the southeast of Europe. There are many relatively young steppe plant taxa that are characteristic for the western side of Eurasia and whose origin can be found in the coastal areas of the ancient Mediterranean Sea (Tethys) (species of *Stipa*, *Festuca*, *Koeleria*, *Agropyron* s. str., *Astragalus*, *Onosma*, *Salvia*, *Artemisia*, etc.) (Hurka et al. 2019).

In the extreme southeast of Europe, a small part belongs to the Oriental-Turanian region which is mainly situated outside of Europe in the meridional zone (central Anatolian, Armenian, Syrian, Araksian, Hyrcanian, Iranian, and South Turanian provinces) and the submeridional Aralo-Caspian province. Common taxa are geophytic taxa (*Muscari*, *Tulipa*), thorn-cushion taxa (*Acantholimon*), thistle relatives (*Centaurea*, *Cousinia*), and hapaxanthic xerophytes (*Verbascum*).

The Macaronesian-Mediterranean region extends beyond Europe into the northwest African mountain ranges and along the southern coast of the Mediterranean Sea to Egypt, further into the western outskirts of Asia Minor and the Near East (Israel, Lebanon, West Syria, the Mediterranean margins of Turkey). It is marked by elements of evergreen sclerophyllous forests (*Quercus ilex*, *Q. coccifera*, *Q. suber*, *Olea europaea*, *Ceratonia siliqua)* and remains of laurel forest vegetation (*Laurus nobilis*, *Prunus laurocerasus*, *P. lusitanica*, *Laurus azorica*, etc.) (Fernández-Palacios et al. 2011). Many of these elements extend to the sub-Mediterranean subregion (e.g., *Pistacia*, *Punica*, *Arbutus*, *Cistus*, *Rhamnus alaternus*, *Buxus*, *Laurus*, *Prunus laurocerasus*, *Rhododendron ponticum*). This region covers the richest refuge areas for broadleaved deciduous forests in their central and eastern provinces (particularly in mountainous regions such as the Caucasus, northern Anatolia, Balkans, Illyria, and the southeastern Alps). Thermophilous forests of *Quercus pubescens*, *Q. pyrenaica*, *Q. faginea*, *Q. cerris*, *Q. frainetto*, *Q. polycarpa*, *Carpinus orientalis* as well as *Ostrya carpinifolia*, *Fraxinus ornus*, and *Pinus nigra* predominate over large areas. Evergreen vegetation, most genera are of Mediterranean origin, occur in mild winter coastal areas of this subregion. The Atlantic islands (Canaries, Madeira, and the Azores), never been connected to the mainland, were colonized mainly by species from the Mediterranean region. Therefore, they are associated with the Mediterranean region as the Macaronesian subregion characterized by many plants of laurel forests (Press and Short 2016).

The endemic flora of the islands reflects their considerable age (Axelrod 1975). Fossils of leaves and fruits found in many places in the Mediterranean region and South Russia (Barcelona, Rhone Valley, S. Italy, Godanski Pass, etc.) are of plants identical to species now found only in the Canary Islands and Madeira. These fossils of plants such as the Dragon tree, the Canarian laurels, and many of the Canarian ferns date from the Miocene and Pliocene periods of the Tertiary Epoch and are up to 20 million years old (Cronk 1992). During this period the Mediterranean region formed part of the basin of an ancient ocean, the Tethys Sea, which separated Europe from Africa. On the margins of this subtropical sea, the vegetation must have been very similar in compositions and appearance to the laurel forest communities of the present-day Canary Islands (Bramwell and Bramwell 1974; Del Arco Aguilar et al. 2010). The vegetation dynamics on the European and African continents since the islands emerged provide the key to their main sources of flora (Del Arco Aquilar and Delgado 2018).

The Late-Glacial and Holocene Vegetation History of Europe (Lang et al. 2003)

Introduction

Since the end of the Tertiary, the development of the present-day vegetation cover of Europe has been decisively influenced by pronounced climatic changes. From the beginning of the Quaternary, several glacial periods alternating with warmer interglacials occurred, apparently as a consequence of periodic changes in the Earth's orbit. The repeated climatic changes led to a successive shift in the distributions of plant species and in particular to a gradual loss of woody species within Europe (Lang et al. 2003).

In the development of the present vegetation, the Late-Glacial (15,000 to 10,000 BP) and the Post Glacial or Holocene (10,000 BP to present) eras have been decisive periods. All age data characterized with BP relate to conventional radiocarbon years before present, i.e., to uncalibrated carbon-14 data. So these date specifications are 500–1,000 years too young in the middle and early Holocene compared to calendar years, which are given in years BC (Before Christ) or AD (Anno Domini). From the end of the Last Glacial Epoch/Period, when most of Europe was unforested (Fig. 1), woodland gradually developed to become the dominant natural formation of today. As in the preceding interglacials, vegetation development was determined by natural climatic changes, particularly temperature, and the development of soils. Subsequently this led to a relatively rapid migration of plants.

The development of the present real vegetation ("aktuelle Vegetation") during the Holocene, however, was to a great extent influenced also by a second factor, which was absent during the interglacials, which is the constant increasing influence of humans from the beginning of the Neolithic period. Over the period from 10,000 to 5,000 BP, non-nomadic agricultural communities spread from the Near East to Southeast Europe and then to the northwest, north, and northeast. This led to considerable changes and displacement of vegetation especially of woodland and resulted in a change from a "natural" landscape to a "cultural" landscape. That implies changes in mesoclimates, transformations of forest soils into agricultural ones, changes in the water regimes, and the appearance of erosion. Regional effects were secondary plant migration, changes in vegetation zonations (both horizontal and altitudinal), the development of new agricultural and synanthropic ecosystems, steppes and deserts (aridization and desertification), as well as increased paludification and expansion of peatlands, especially in the uplands and in the north (Lang 1994).

Because the Late-Glacial and Holocene vegetation history of Europe is so diverse, the continent ought not to be dealt with as a whole. Instead vegetation development should be considered in several individual and smaller regions. The zonation of the present natural vegetation can be used as a basis for this purpose (Fig. 5). This includes a sequence of zones extending from north to south, i.e., from arctic tundras to boreal coniferous forests, temperate (nemoral) deciduous forests, Mediterranean sclerophyllous forests, and Pontic-Caspian steppes and semideserts. In addition, three major forest regions can be distinguished from west to east corresponding to their decreasing oceanicity and increasing continentality; the Colchis and Caucasus region is also included in the east of Europe. Naturally, most of these areas are not heterogeneous in themselves due to differences in, e.g., relief (valleys and mountain ranges) and substratum (limestone and silicates), with their own individual influence on the development of vegetation. As far as possible, this fact shall be considered in the following description. However, we are not able to discuss the changes in azonal vegetation, because of the limited scope of this text. Regarding the order of the regions described, we shall start in the south and southeast, effectively following the Late-Glacial and Holocene spreading of forests from the south/southeast to the north/northwest, as well as the more or less similarly directed advance of permanent agricultural settlement throughout Europe (Lang et al. 2003).

The present distribution of plant communities and their zoning, as shown in the vegetation map of Europe, should be considered the result of a long and extremely variable development. Inducing and controlling factors during the course of the Holocene were not just the continuous changes in environmental conditions but also the increasing scale of human intervention. Probably there is not a single part of Europe that has been untouched by human activities. Even the highest mountain peaks or arctic tundra have been affected, whether it be directly or indirectly as a result of air pollution or other emissions (Lang et al. 2003).

Of course, present vegetation patterns do not represent an end of the developments described, and they will indeed continue to develop in the future, irrespective of whether such changes will be shaped by natural or artificial forces. Following Lang et al. (2003), the vegetation history will be described for every region.

Mediterranean Zone

The Late-Glacial and Holocene vegetation history of the Mediterranean area can be differentiated from the northern parts of Europe in main features. Notably, by the presence of evergreen forests already since the Tertiary, by the survival of many deciduous forests in glacial refugia during the Pleistocene ice ages, and by the intense and long-lasting agricultural influence on vegetation during the Holocene, three areas can be distinguished, namely, the western, central, and eastern region.

Western Mediterranean Region

This region covers the major part of the Iberian Peninsula and the southern coast of France, which represent areas with variable relief, climate, and flora. Relatively little information is available regarding the vegetation history (Sainz Ollero and van Staalduinen 2012).

Artemisia steppes with scattered occurrences of *Pinus* were common during the early Late-Glacial. In the more recent Late-Glacial and at the start of the Holocene, deciduous *Quercus* forests spread out, with *Betula* at higher altitudes. *Corylus* was always of less importance. During the Pleistocene ice ages, both deciduous and evergreen *Quercus* species were presumably able to survive in glacial refugia in this region. Evergreen *Quercus* and other evergreen trees were already present near the coast in the south and southwest during the Late-Glacial and in other areas from about 8,000 BP. Presumably human influence like grazing, fire, and consequent soil erosion encouraged the expansion of evergreen tree and shrub species inland to the north. *Abies pinsapo*, an endemic species, most likely existed in the south Spanish mountain range in a glacial refugium. In the first half of the Holocene, *Fagus* and *Abies alba* spread out to the south of France and the Pyrenees from the east.

The oldest Neolithic cultures, which date back to about 7,500 BP, have been confirmed in Southern Spain and the south of France. However, in other parts of the region, the influence of grazing on vegetation has also been observed from this time on. In the montane plateaux of Portugal and Spain, pollen analyses have revealed that intense cultivation of cereals and *Olea* was already practiced from about 3,000 BP which produced major landscape and vegetation changes, e.g., *Erica arborea* expanded as a result of increasing grazing (Lang 1994; Kosmas et al. 2002).

The land degradation in the Mediterranean, for example, in Spain results from various factors including climatic variations and human activities (Thornes 2002). These definitions of "desertification" include the degradation of soil, vegetation, as well as ecological processes that operate within the system (Millána et al. 2005; D'Odorico et al. 2013).

Central Mediterranean Region

Corsica, Sardinia, and Sicily, the greater part of the Apennine peninsula, and the narrow eastern margin of the Adriatic coast belong to this region. As other parts of the Mediterranean, it is characterized by a very varied relief (with peaks of 3,270 m in Sicily and 2,710 m in Corsica) and a correspondingly varied climate and vegetation. Local mountain glaciation during the ice ages had scarcely any direct influence on the vegetation development.

During the early Late-Glacial, the region was characterized by extensive open treeless vegetation. The *Artemisia* dominated steppes were presumably scattered with small stands of deciduous trees and also *Abies,* particularly in the southernmost parts of the Apennine peninsula. The region is considered to be an important glacial refugium. In the younger Late-Glacial, a scrub vegetation (open woodland) developed, with different deciduous broadleaved trees (*Ulmus*, *Tilia*, *Fagus*, *Carpinus*, *Corylus*) but mainly deciduous *Quercus* species, as well as various *Pinus* species and sporadic *Abies*.

In the early Holocene, mixed deciduous *Quercus* forests covered the lower altitudes, while *Fagus-Abies* forests were predominant in the mountains. In the second half of the Holocene, evergreen vegetation gradually began to replace the deciduous oak forests, an effect of more and more activities of the Neolithic populations, grazing by domesticated animals as well as fire and increased soil erosion. In Southern Italy, archaeologists have dated the beginning of the Neolithic Age to about 7,000 BP (6,300 BC) and the start of cereal cultivation at about 6,500 BP. However, the extensive deforestation and intensive arable farming occurred later at about 3,000 BP. Islands such as Corsica have, in some respects, run through slightly differing vegetation histories: *Pinus nigra* subsp. *laricio* was locally present since about 12,000 BP; *Erica arborea* has prevailed since the first half of the Holocene. Particularly in the central Mediterranean, great changes in vegetation occurred during the period of the Roman Empire when numerous secondary plant communities originated (Lang et al. 2003).

Eastern Mediterranean Region

The Eastern Mediterranean region is made up of the southern part of the Balkan Peninsula and the Aegean Islands region. Diverse climatic and floristic conditions occur due to the large variations in altitude in this area (Mount Olympus 2,911 m). The local mountain glaciation during the Pleistocene ice ages had hardly any influence on the development of the vegetation.

In the early Late-Glacial period at middle altitudes, steppes with *Artemisia* and *Ephedra* among other species dominated with probably sporadic small trees and/or shrubs. The region was a glacial refugia for many European forest trees, presumably, e.g., for *Abies*, *Acer*, *Carpinus betulus*, *Ostrya*, *Fagus*, *Fraxinus*, *Ulmus*, *Tilia*, *Corylus*, *Olea*, and *Quercus* (both deciduous and evergreen species). After that, the deciduous forests expanded (deciduous *Quercus* species, *Ulmus*, *Tilia*, *Ostrya*), while coniferous forests with *Pinus* spread locally. Forests with *Fagus* and *Abies* occurred sporadically at higher altitudes. In the first half of the Holocene, mixed deciduous forests covered the region up to high altitudes. After about 6,500 BP, these were accompanied by *Ostrya/Carpinus orientalis*.

Settlement began very early in the lower parts of the region with mixed deciduous oak forests. From about 10,000 to 9,000 BP, first agricultural grazing cultures, keeping goats and sheep, are known. Between 9,000 and 8,000 BP, archaeologists dated the first cereal cultivation. According to palynological data, from about c. 3,000 BP, which corresponds approximately to the Hellenistic era, intense agriculture began. This includes fruit farming (cereals, *Olea*, *Vitis*, *Juglans*, *Castanea*) connected with widespread timber-felling. A secondary expansion of evergreen trees (*Quercus ilex*, *Pistacia*, etc.) was initiated by clearing, grazing, and fire with subsequent soil erosion. In limestone areas severe soil erosion led to karst formations. Apparently natural xeric plant communities formed as a result of secondary migrations of numerous plants, and shifts in vertical vegetation belts occurred. Only a few mountain areas kept their natural character (Lang et al. 2003).

Temperate Zone

The temperate zone includes the majority of Western, Central, and Eastern Europe as well as mountain ranges with deciduous forests in this area. In the Late-Glacial and early Holocene, the vegetation development of these areas is characterized by a succession from forest-free vegetation to *Betula-Pinus* stands, and a subsequent expansion of mixed broadleaf woodland followed, sometimes with a contribution from various conifers. During the entire Pleistocene Epoch, in spite of climatic alternations, soils and their vegetation cover remained, although subject to many changes in the non-glaciated periglacial areas. By contrast, in the glaciated areas, the development of soils and vegetation began repeatedly with initial phases on bare, vegetation-free ground. In the temperate zone, human settlement started at the latest at around 7,500 BP from the southeast and went on until the Middle Ages. Related human activities resulted in an extensive deforestation and changes in vegetation, as described below in further detail. Thus, eight regions can be distinguished including separate mountain ranges, characterized by specific vegetation sequences (Lang et al. 2003).

Balkan Region

The region covers a considerable part of Southeast Europe, the mountain ranges of the Balkan Peninsula (the Dinaric Alps, Rhodopes, etc), and the Danubian lowlands. During the last glacial period, local glaciers occurred. As the Eastern Mediterranean Zone (Mc), the region particularly served as an important glacial refugia for many Central European trees, e.g., for *Abies alba*, *Acer*, *Carpinus betulus*, *Fagus*, *Fraxinus excelsior*, *Ostrya*, *Tilia*, and *Ulmus*, and some deciduous *Quercus* species. Despite this importance, just a few studies have been made on the vegetation history of this area.

In the Late-Glacial period, montane areas were covered with open steppe vegetation, with sporadic occurrences of *P. sylvestris*, *P. nigra*, *P. heldreichii*, etc. Refugial stands of Central European deciduous trees during the glacial as mentioned above were most probably present at low altitudes. In the Holocene, above all, a strong expansion of mixed *Quercus* forests with *Carpinus*, *Ostrya*, *Ulmus*, and *Tilia* took place. In the middle Holocene these forests reached much higher altitudes than they do today. The mass expansion of *Abies* occurred between 4,000 and 6,000 BP, while that of *Fagus* started later, after about 4,000 BP.

Similar to the Eastern Mediterranean region, agriculture started very early in this region. According to archaeological findings, Neolithic settlements date back to about 7,500 BP. After 3,500 BP, considerable mountainous areas were deforested by the ancient Greeks for sheep breeding. This led to soil erosion, karst formation, and the secondary expansion of many plants (Lang et al. 2003).

Central European Region

This region consists of lowlands in the north and at the southern edge of the Alps as well as numerous low mountain ranges, including the Harz (1,142 m maximum), the Sudetic Mountains (1,602 m), the Erzgebirge (1,244 m), the Bavarian-Bohemian Forest (1,456 m), the Black Forest (1,493 m), the Vosges (1,426 m), the Massif Central (1,886 m), and the North Apennines (2,165 m). The Pyrenees (Tf), the Alps (Tg), and the Carpathian Mountains (Th) are described individually later. During the last glacial period, the northern parts of the Central European Region were covered by spurs extending from the northern ice sheet. The Alpine glaciers extended far over the foothills, but the higher low mountain ranges showed only local glaciation. During the early Late-Glacial period, all these glaciers started melting. However, most of the region remained ice-free during the whole Pleistocene. Those periglacial areas that remained ice-free during the penultimate ice age (Riss glaciation) permitted the continuous development of soils and vegetation for a considerable period of time.

At low altitudes the progress of reforestation of open steppe-tundra landscapes had already started in the Late-Glacial: in the south at around 12,500 BP during the Bølling and in the north, a little later, during the Allerød period. *Betula* dominated the vegetation in the northwest and west and *Pinus sylvestris* in the southwest, southeast, and east. In the first half of the Holocene, vegetation development, especially in the Boreal, was marked by the mass expansion of *Corylus* particularly in the west and then later in the Atlantic by the spreading of *Quercus*, *Ulmus*, *Tilia* and other deciduous trees. In the second half of the Holocene, *Fagus* spread continuously from the southeast to the north and northwest. In the southeast and east, it invaded *Picea* stands, and in the west, it replaced some *Quercus* mixed forests. In the south, *Fagus* was accompanied and sometimes even replaced by *Abies alba*. In the Holocene, *Picea* abies migrated gradually from east to west into the low mountain ranges north of the Alps. But it was never as frequent as in the high Carpathians or the interior of the Alps.

The first isolated forest clearings in loess areas of early settled landscapes mark the start of the Neolithic period, which dates back to 7,000 BP in the lowlands, the southeast, and the center and in the north to around 5,000 BP. From the Bronze Age and the Iron Age to the Middle Ages, human interference grew more and more. Low mountain ranges are the most recently settled areas, and the beginnings of agriculture often date back only to the Middle Ages. Extensive human impact as clearing, grazing, and crop farming often caused secondary plant migrations, modified the natural boundaries of vertical vegetation belts, and led to the formation of xerothermic communities on flat land (Bazha et al. 2012). In lowlands agricultural activities locally resulted in soil waterlogging, at higher elevations and at higher altitudes in the expansion of open damp vegetation. The formation of meadows, pastures, and other substitute communities of cultural landscapes can be traced back to these processes (Lang et al. 2003).

Atlantic Region

This region includes most of the British Isles as well as coastal strips of the European mainland from southwest Norway to the north of Portugal. The area is made up of mountains as well as lowlands with accordingly different present vegetation and different vegetation history.

During the last glacial period, the highlands of Scotland and Northern England, Wales, and Ireland were mostly covered under a continuous ice sheet. By the end of the Late-Glacial, this had melted to relatively small remnants. In the Late-Glacial, the ice-free areas were settled by open pioneer vegetation with features of dwarf shrub tundra with *Salix herbacea*, *Betula nana*, *Juniperus*, etc. During the Allerød oscillation, birches trees may have been present occasionally. In the Preboreal *Betula* and *Pinus sylvestris* became the dominant tree species. *Corylus* expanded from 9,000 BP. During the Atlantic scattered mixed deciduous *Quercus* forests developed with *Quercus* and *Ulmus*. *Alnus* also spread rapidly during this period. Simultaneously a gradual decrease in forestation began with increasing paludification and the formation of blanket bogs at the same time. The beginning of intensive human interference with the vegetation may have occurred between the first Neolithic settlement around 5,000 BP and the Bronze Age. From around 3,500 BP, agricultural activities grew more intense. Especially, cattle and sheep grazing promoted the formation of blanket bogs and led to the development of heaths and grasslands.

In the lowlands of England, Ireland, and the Atlantic coastal strips, the sequence of vegetation development was very similar in the Late-Glacial and early Holocene. During the Boreal and Atlantic periods, *Quercus* mixed forests are dominated, and *Alnus* expanded rapidly in this area. A specific incident in the vegetation history of this region is the so-called Elm decline at around 5,000 BP. *Fagus* arrived in the region between 3,000 and 1,000 BP, but could not establish itself as an important component of the vegetation. In Ireland, Northern England, and Scotland, *Fagus* is naturally quite absent.

As mentioned above, anthropogenic transformations of the vegetation began during the Neolithic period at around 5,000 BP in this part of Europe. The agricultural activities of the early settlers consisted basically of arable farming and grazing. Typical early settlements were scattered local clearings within a forest landscape, the so-called Landnam. More extensive deforestation followed from the Bronze Age onwards and reached its maximum during Roman times and the Middle Ages (Lang et al. 2003).

Eastern European Region

The west of Ukraine, southern Belarus, and parts of central Russia belong to this region. The characteristic landscape is flat or slightly hilly lowland. Most of this area remained ice-free during the last glacial period. Huntley and Birks (1983) assumed that *Acer*, *Ulmus*, *Tilia*, and maybe also *Corylus* found glacial refugia in this region.

In the older Late-Glacial, typical vegetation in this region was steppes. During the Allerød they were replaced by *Pinus-Betula* forests. In the early Holocene deciduous trees, *Quercus*, *Acer*, *Tilia*, *Ulmus*, and *Carpinus* spread into the pine-dominated forests. Mixed deciduous *Quercus* forests with *Corylus* spread out particularly in the west in the middle Holocene. At the same time, expansion of *Alnus* started, and *Picea* spread from the northeast. *Fagus* and *Abies* never reached this area.

In this region, according to archaeological data, the Neolithic period started at around 6,000 BP, though Ukrainian and Russian pollen diagrams have provided almost no evidence for any early agricultural cultures. The oldest available signs of agriculture from pollen analyses, the record of cereals and *Centaurea cyanus*, date from the period between 2,500 and 2,000 BP. *Fagopyrum* pollen grains have been verified from around 900 BP in Kulikovo Pole, south of Moscow (Khotinskij, unpublished results). Belarus and northwest Ukraine were settled by hunters and fishermen; this kind of land use did not particularly change the natural landscape. The continuity of farming cultures was repeatedly interrupted because of repeated attacks by belligerent nomadic peoples from Asia, so that settlement discontinuities were characteristic. This applies to the Pontic Zone (P) too (Lang et al. 2003).

Hemiboreal Region

This region includes Southern Scandinavia, the Baltic States, and the extreme southwest of Finland, as well as northern Belarus with wedges extending eastwards into central Russia and is located at the southern margin of the boreal zone. Large parts of this region were under a continuous ice sheet during the last glacial period, but the whole region had already become ice-free by the end of the Late-Glacial.

At the beginning of the Holocene, *Pinus sylvestris* and *Betula* sect. *Albae* were the predominant trees. From glacial refugia located in the east, *Picea* abies gradually spread to the west reaching the Baltic by around 5,000 BP and Southern Sweden by around 2,000 BP. In the early Holocene, several deciduous broadleaved trees, such as *Corylus*, *Quercus*, *Tilia*, and *Ulmus*, immigrated from the south. Particularly during the Atlantic Period, forests consisted of these deciduous trees and conifers. *Carpinus* reached the region only after 5,000 BP.

This region has been settled more or less continuously since the Bronze Age (at the latest). Local hunters and fishermen lived in the region with little influence on the vegetation. For the south of Sweden, pollen diagrams show first traces of clearing and farming between 4,000 and 3,500 BP. In the Baltic States, pollen diagrams record first agricultural activities considerably later from around 1,000 BP. Though archaeological findings date cereal farming back to a period between 3,800 and 3,000 BP, the absence of pollen records indicates that it probably was not widespread at this time. Major agricultural expansion leading to severe deforestation and changes in landscape only started in the Middle Ages (Lang et al. 2003).

The Carpathian Region

The major part of the Carpathian Region remained ice-free during the Pleistocene. Only local glaciers covered the highest mountains (High Tatras 2,663 m and South Carpathian Mountains 2,544 m) during the last glacial period. By the early Late-Glacial period, they had already melted. The highly variable topography leads to diverse climate, soil, and vegetation conditions, respectively. At low altitudes, in the intermontane basins of the west, Carpathian Mountains glacial refugia of conifers *Larix*, *Pinus cembra*, and *Picea* forests were situated. In the Late-Glacial, *Pinus* was the main woody plant in the entire region. Special conditions existed in the intermontane basins of the Western Carpathian Mountains. There is evidence of open forests with *Pinus cembra*, *P. sylvestris*, and possibly also *P. mugo*, *Larix decidua*, *Picea abies*, and *Juniperus* in this area. After about 4,000 BP, the earlier *Pinus-Picea* stands were replaced by mixed beech forests with *Abies* and *Picea* at mid-altitudes. *Picea* kept its dominance only in the altimontane belt, especially in the High Tatras and their foothills. Deciduous *Quercus* forests mixed with *Ulmus*, *Tilia*, and *Corylus*, developed at low altitudes already during the Preboreal and Boreal in the southern and eastern Carpathian Mountains, but in the Western Carpathian Mountains, this shift occurred later in the Atlantic Period. *Carpinus* spread to the northwest from the southeast of the Carpathians during the early and middle Holocene.

The lowlands of the Carpathian Mountains and their foothills were inhabited by Neolithic settlers already around 7,000 BP. Agricultural exploitation increased from the Bronze Age onward. The most intensive deforestation and clearance occurred particularly in the Middle Ages and during the Wallachian colonization in the sixteenth and seventeenth centuries. The Wallachian livestock farming, with summer grazing in the mountains, has greatly influenced the contemporary landscape and vegetation and led to an anthropogenic lowering of the natural alpine tree line in many places.

Alpine Region

The Alps were covered by an enormous ice sheet during the last as well as preceding glacial periods. Only the southwest and the extreme east of the mountain range was not covered by glaciers, and some few nunataks (ice-free mountain peaks within glaciated regions) existed at the northern and southern margins of the Alps. During the Late-Glacial, the glacier masses receded in the inner alpine area; today, only a few glaciers remain. Considerable variability in both altitude and geology determines a great ecological variety in this region.

In the northeast, eastern, and southeast margins of the Alps, in local climatically favorable sites, *Picea abies* may have survived through the last ice age. In the younger Late-Glacial, *Pinus sylvestris*, *P. cembra*, *Larix*, and *Juniperus* dominated at lower altitudes, while at higher altitudes, open pioneer vegetation, alpine tundra, and grasslands were found. In the subsequent Holocene, vegetation developed differently in the marginal and in the Central Alps. In the Central Alps, *Picea abies* spread out from the east and southeast, mostly between 11,000 and 5,000 BP, without reaching the French southwestern Alps. During the Atlantic period, the tree line, with open *Pinus cembra-Larix-Juniperus* forests, may have been a maximum of 200 m higher than it is today. At the tree line, *Alnus alnobetula* was also an essential element of vegetation from about 5,000 BP in many places. In the marginal Alps and their foothills, *Corylus* occurred in great quantities at the lower elevations in the Boreal and Atlantic periods. Then mixed deciduous *Quercus* forests developed, into which *Carpinus* migrated from the southeast from about 6000 BP. Later mixed forests with *Fagus*, *Abies*, and sometimes *Picea* developed between 7,000 and 5,000 BP.

Neolithic people colonized lower parts of the Alps and their foothills from about 7,000 BP. Animal farming and intense grazing in the mountains can be traced back to Roman times. At the same time, the earlier local deforestation reached greater dimensions in the wide alpine valleys and basins. Present-day landscape was formed in the Middle Ages. The anthropogenic deforestation and grazing encouraged a secondary expansion of the alpine flora and led to a general lowering of the tree line (Lang et al. 2003).

Pyrenean Region

This high mountain range in the southwest of Europe has peaks of over 3,400 m. As in the Alps and the Carpathian Mountains, large differences in altitude produce remarkable climatic and vegetational variations which also show in the vegetation history. During the early Late-Glacial period, glaciation of the last ice age already melted away rapidly.

Already in the early Late-Glacial, the initially dominating steppe landscape turned into *Pinus-Betula* forests at lower (300–1,100 m) and middle altitudes (1,100–1,800 m). At the beginning of the Holocene, mixed deciduous *Quercus* forests with *Corylus* and *Ulmus*, and later with *Tilia*, started to develop. Between 8,000 and 5,000 BP, *Abies* expanded from east to west, followed by *Fagus* between 4,500 and 4,000 BP. *Fagus-Abies* mixed forests originated at middle altitudes. *Picea*, a characteristic component in the vegetation of the Carpathian Mountains and the Alps, did not reach the Pyrenees. Throughout the entire Holocene *Pinus* predominated at higher altitudes (about 1,800–2,000 m). It can be supposed that significant variations occurred in the Alpine tree line. Mixed deciduous *Quercus* forests may have advanced to higher altitudes in the middle Holocene than they do today.

In the Pyrenean Region, pasturing has been recorded from about 4,000 BP, even though no major deforestation occurred before the Middle Ages. At lower altitudes and in the foothills, the oldest signs of arable farming could be dated back to about 5,000 BP, at middle altitudes; however, arable farming did not start before the Middle Ages (Lang et al. 2003).

The Colchis and Caucasus Zone

This region can be divided into four subregions, the Great Caucasus, the Lesser Caucasus, the plateau lying between them in the east with the river valley of the Kura, and the Colchis in the west. However, they are not marked on the map in Fig. 5, yet the four subregions are described in the following text. During the coldest periods of the Pleistocene, the high mountain ranges were glaciated. Today, however, only small glaciers are left over at the highest altitudes in the Great Caucasus. Characteristic of the Colchis and Caucasus are considerable differences in altitude (Elbrus 5,633 m), diverse geological conditions, and the location at the juncture of several Eurasian phytogeographical zones. These features have all had a major influence on the vegetation-historical development and effects on the current status, respectively. During the Quaternary, migration, expansion, and replacement of plant species to and from other areas of Europe were limited because of various barriers and the lack of any suitable migration routes. Migration and exchange were only possible via the Crimea or northern Anatolia, for example. Because of this isolation, the present flora and vegetation are highly specific, and the Caucasus represents an important refugial area and center of speciation. Within Europe it has the highest number of endemic plant taxa.

During the entire Holocene, the ice-free parts of the Great Caucasus were a woodland mountain range. After the decline of mountainous glaciation, subalpine and alpine tundra communities developed in these areas primarily. At middle altitudes the tundra was already replaced by *Pinus kochiana* and *Betula-Salix* forests in the early Holocene. Then during the middle Holocene, extensive *Fagus sylvatica subsp. orientalis, Abies nordmanniana*, and *Picea orientalis* forests developed, which still exist today.

During the Holocene, unlike the Great Caucasus, the Lesser Caucasus seemed to be covered with herb-rich, xerophytic steppe, and forest steppe, that is largely forest-free. Only small areas were wooded, and these developed similar to those of the Great Caucasus.

In the periglacial areas between the two main mountain ranges, including the Kura river valley, forests and forest steppe have been recorded with scattered relict stands of deciduous broadleaved forests, for the last glacial period. From the early Holocene, the forests expanded and then achieved dominance during the middle Holocene. The forests were made up of *Juglans*, *Quercus*, *Castanea*, *Ulmus*, *Carpinus*, *Acer*, *Fagus*, *Pterocarya* and with some *Picea* to be found. At the same time, extensive floodplain forests developed in the valleys.

In the Colchis, the eastern Caucasian lowlands, and in Armenia, very early arable farming has been confirmed. This type of land use probably spread from Mesopotamia. The large mountain ranges were populated by hunters and shepherds. The Caucasian countries were invaded or traversed by neighboring peoples on many occasions and became especially important as a metallurgical center. The present severely deforested, agriculturally and horticulturally influenced landscape developed, in large part, during the early feudal periods of Transcaucasian history, between the third and ninth centuries AD (Lang et al. 2003).

The Pontic-Caspian Zone

The Pontic Zone of Europe is a flat lowland which forms the western spur of the extensive Eurasian belt of wooded steppe, steppe, and semidesert. Vast aeolian Quaternary sediments (loess with chernozem soils and sands) and alluvial sediments are characteristic. The area remained ice-free during the Quaternary Period.

Only a few pollen analysis have been published from this zone so far, so a subdivision into regions is not possible yet. Today the region represents a floristically modified remnant of the cold steppe and *Artemisia* semidesert areas that was once widespread throughout Europe during the glacial period. The vegetation development in this zone was characterized by the dominance of steppe communities with *Artemisia*, Poaceae, and Chenopodiaceae during the entire Holocene. The non-tree pollen (NTP) values always lie between 70 and 90% of the total pollen count. From 6,000 BP pollen of deciduous trees as *Quercus* and *Tilia* has been verified sporadically, as has pollen from *Pinus* and *Betula*. But rates of these last-named species are so low that the presence of these trees could be considered as questionable. The concept of glacial refugia of deciduous broadleaved trees in river valleys of the Southern Ukraine and Russia has not been confirmed so far. In Eastern Romania and Eastern Bulgaria, the extreme southwest part of the zone, a temporary penetration of forest steppe into the steppe region, was observed in the middle Holocene. The species found were deciduous *Quercus*, *Ulmus*, *Tilia*, *Carpinus*, and *Corylus* (Hurka et al. 2019).

In pre-agricultural times, the population of the ancient settled areas in Southern Ukraine and Russia were hunters and fishermen. They lived near the rivers and did not affect or change the steppe grassland. First signs of arable farming date back to about 7,000 BP on the Bulgarian Black Sea coast and to about 4,000 BP in the upper Don region. Initially the conversion of steppe into arable land was moderate. The main reason for this was repeated invasions of Asiatic nomadic tribes which interrupted and impeded any agricultural activity (Smelansky and Tishkov 2012). Large-scale agricultural exploitation of the steppe only began in the Middle Ages and was almost entirely destroyed during the final decades of the former Soviet Union (Lang et al. 2003).

The Boreal Zone

The circumpolar Boreal region extends all the way across the north of the Northern Hemisphere through the Eurasian and the American continent. This zone represents the European part of it. The typical vegetation consists mainly of coniferous forests. During the Pleistocene ice ages, the greater part of the European region was covered by the extensive, contiguous Scandinavian ice sheet. This ice sheet melted away and left only small relicts in the Scandinavian Mountains and Iceland during the interglacial periods. The extensive and long-term ice cover has strongly influenced surface forms, soils, and vegetation and certainly also the vegetation development during the Late-Glacial and Holocene. Apart from the High-Oceanic Region in the west, *Betula* and *Pinus sylvestris* were the dominant trees during the entire Holocene. Another important tree species in boreal coniferous forests (taiga) was *Picea abies*, although its expansion from east to west took place a little later. Compared to the temperate zone, the human intervention and influence on vegetation is significantly lower, and arable farming in this region is of minor importance.

The Boreorussian Region

The region covers the larger part of Northern Russia east of Lake Ladoga and south of the Kola Peninsula. During the last glacial, the western part was covered by inland ice and became ice-free during the Late-Glacial. The eastern part remained completely ice-free.

During the Late-Glacial and Holocene, the dominant tree species were *Betula* and *Pinus sylvestris* and slowly spreading from the east *Picea abies*. Presumably in the ice-free areas of the western foothills of the Urals, *Picea* had a major glacial refugia, from which this conifer spread over almost the entire Fennoscandia Region during the Holocene. Deciduous broadleaved trees as *Corylus*, *Tilia*, and *Ulmus* and sometimes also *Quercus* could, if at all, be found only during the Atlantic Period and sporadically in southern parts of this region. The polar tree limit shifted clearly northwards in the marginal parts of present forest tundra in the middle Holocene and moved to the south again after 5,000 BP with *Betula* sect. *Nanae* (dwarf birches) and *Alnus* sect. *Alnobetula* becoming more important.

Until modern times, human influence on vegetation is hardly noticeable and affected probably only restricted areas. Signs of local clearings have been traceable by pollen analyses only since the Middle Ages at the earliest, but there are no signs of any arable farming. Major activities of the local population were hunting, fishing, and grazing at least since the Bronze Age onwards (Lang et al. 2003).

Fennoscandia Region

This area covers the main part of Fennoscandia including the Scandinavian Mountains, apart from the south and a narrow coastal strip in the west and the north. During the last glacial period, the complete area was covered by inland ice. The enormous ice masses withdrew only during the early Holocene, between 9,000 and 8,000 BP, leaving local mountain glaciers that corresponded approximately to their present pattern of occurrence. In Northern Europe the melting of the ice sheet led to several modifications in the coastline, which also had an influence on the development of the vegetation.

After the retreating of the ice cover, *Pinus sylvestris* and *Betula* migrated rapidly to the ice-freed areas forming the taiga-type coniferous forest. As in the Boreorussian Region, *Betula* was common during the colder periods of the Late-Glacial and the Holocene, while *Pinus* was dominant in the warmer periods. *Picea* spread from its glacial refugia to the west and southwest. The species reached Eastern Karelia around 7,500 BP, while central Norway was first reached around 1,000 BP. Deciduous broadleaved trees (*Corylus*, *Ulmus*) advanced only temporarily into the southernmost parts of the region during the Atlantic Period; *Alnus* arrived between 8,000 and 7,000 BP. In the Scandinavian Mountains, at higher altitudes, the initial mountain tundra was replaced by *Betula pubescens* (s. l.) forests. As in the polar region, the alpine tree line was also subject to changes, at the end of the Boreal and during the Atlantic periods, it may have been situated about 150–200 m higher and many kilometers further north than today.

According to archaeological findings, human settlement and agriculture occurred in southern areas since about 3,500 BP, i.e., since the Bronze Age. The forest character of the landscape remained and was only locally interrupted by clearing, grazing, and hay making. Sporadic signs of arable farming have been found from the Middle Ages onwards (Lang et al. 2003).

Boreoatlantic Region

This region is made up of Northern Scotland including the Outer Hebrides, the Norwegian coast, the islands in the North Atlantic, and Iceland. During the last glacial period, the entire region was covered with ice but became ice-free from the beginning of the Holocene. Only on Iceland, extensive glaciers still remain today.

The Holocene vegetation development of the region was strongly affected by *Betula*. In the greater part of Scotland and the North Atlantic islands, open scrub tundra with *Betula nana*, *Juniperus*, and *Calluna* predominated; forest apparently did not develop during the Holocene. Trees such as *Pinus sylvestris*, *Quercus*, and *Corylus* spread only locally into the region from the Boreal Period onwards. On the Norwegian coast *Betula pubescens* (s. l.) dominated in the vegetation development from about 12,000 BP.

Pinus sylvestris may only have appeared occasionally, if at all. On Iceland the ice-free areas were colonized by open tundra and scrub tundra. Especially in the south of the island and only during the Atlantic period, taiga-like birch stands expanded. However, in consequence of woodcutting, fire, and grazing, these stands disappeared more and more from the Subboreal period onwards, especially after the Vikings colonized Iceland around about 1,100 BP.

The population practiced sheep farming in most parts of the region (the Norwegian coast, Shetland Islands, and the Faeroe Islands). The grazing sheep promoted the development of grasslands and heaths. Compared to other parts of Europe, anthropogenic impact on the vegetation remained quite limited (Lang et al. 2003).

Arctic Zone

This zone describes the northernmost part of Fennoscandia and Russia and the Arctic islands in the Arctic Ocean. During the last glacial period, the western part was covered by ice which had already melted away by 13,000 BP.

During the Late-Glacial, different vegetation types spread across the ice-freed soils, freeze-thaw polygonal tundra, and then lichen-moss tundra and eventually dwarf shrub and open birch tundra. In the mainland areas, during the Boreal Period, birch forest tundra was present. In the subsequent Atlantic period between 7,000 and 5,000 BP, the tundra was replaced by taiga with *Picea abies* predominating. During this period *Pinus sylvestris* occurred as well as *Pinus sibirica* in the east, and most of today's subarctic bogs development started. After 4,000 BP the taiga withdrew southward, and scrub and birch forest tundra spread again. By the end of the Subboreal period, approximately 2,500 BP, the present vegetation patterns developed (Lang et al. 2003).

Until the very recent past continuous settlements did not exist in this region. Human influence on vegetation has only occurred over the last decades as a result of geological prospecting and mining activity, etc., mainly in the form of mechanical damage.

References

Acosta J, Uchupi E, Muñoz A, Herranz P, Palomo C, Ballesteros M, ZEE Working Group (2005) Geologic evolution of the Canarian Islands of Lanzarote, Fuerteventura, Gran Canaria and La Gomera and comparison of landslides at these islands with those at Tenerife, La Palma and El Hierro. Mar Geophys Res 24:1–40

Afonso L (1988) Geografia de Canarias, 2nd edn. Interinsular Canaria, Santa Cruz de Tenerife

Andersen BG, Borns HW Jr (1994) The ice age world. Scandinavian University Press, Oslo, 2089p

Axelrod DI (1975) Evolution and biogeography of Madrean-Tethyan Sclerophyll vegetation. Ann Mo Bot Gard 62:280–334

Bazha SN, Gunin PD, Danzhalova EV, Drobyshev YI, Prishcepa AV (2012) Pastoral degradation of steppe ecosystems in Central Mongolia. In: Werger MJA, van Staalduinen MA (eds) Eurasian steppes. Ecological problems and livelihoods in a changing world. Springer, Dordrecht, pp 289–319

Bramwell D, Bramwell ZI (1974) Wild flowers of the Canary Islands. Stanley Thornes Ltd, Surrey, 332p

Carracedo JC, Day S, Guillou H, Rodríguez Badiola E, Canas JA, Pérez Torrado FJ (1998) Hotspot volcanism close to a passive continental margin: the Canary Islands. Geol Mag 135:591–604

Crawford RMN (2000) Ecological hazards of oceanic environments. New Phytol 147:257–281

Cronk QCB (1992) Relict floras of Atlantic islands: patterns assessed. Biol J Linn Soc 46:91–103

Cropper TE, Hanna E (2013) An analysis of the climate of Macaronesia, 1865–2012. Int J Climatol 34:604–622

D'Odorico P, Bhattachan A, Davis KF, Ravi S, Runyan CW (2013) Global desertification: drivers and feedbacks. Adv Water Resour 51:326–344

Del Arco Aguilar M-J, González-González R, Garzón-Machado V, Pizarro-Hernández B (2010) Actual and potential natural vegetation on the Canary Islands and its conservation status. Biodivers Conserv 19:3089–3140

Del Arco Aquilar M-J, Rodríguez Delgado O (2018) Vegetation of the Canary Islands. Plant and vegetation, vol 16. Springer International Publishing AG part of Springer Nature, Cham, 429 p

Dias E, Mendes C, Melo C, Pereira D, Elias R (2005) Azores Central Islands vegetation and flora field guide. Quercetea 7:123–173

Fernández-Palacios JM, De Nascimento L, Otto R, Delgado JD, Garcia-del-Rey E, Arévalo JR, Whittaker RJ (2011) A reconstruction of Palaeo-Macaronesia, with particular reference to the long-term biogeography of the Atlantic island laurel forests. J Biogeogr 38:226–246

FID (ed) (1971) Regionalisation of Europe. La Haye (Fédération internationale de documentation)

Geldmacher J, Van den Bogaard P, Hoernle K (2000) The 40Ar/39Ar age dating of the Madeira archipelago and hotspot track. Geochem Geophys Geosyst 1(2):26

Harris MF, Finger FG, Teweles S (1962) Diurnal variation of wind, pressure, and temperature in the troposphere and stratosphere over the Azores. J Atmos Sci 19(2):136–149

Hübl E, Niklfeld H (1973) Über die regionale Differenzierung von Flora und Vegetation in den österreichischen Alpen. Acta bot Acad Sci Hung 19:147–164

Huntley BJ, Birks HJB (1983) An atlas of past and present pollen maps for Europe. Cambridge Univ. Press, Cambridge, 667p

Hurka H, Friesen N, Bernhardt K-G, Neuffer B, Smirnov S, Schmakov A, Blattner F (2019) The Eurasian steppe belt. Status quo, origin and evolutionary history. Turczaninowia, 2019.23.3: 000-000

Jäger EJ (2003) Phytogeographical classification of Europe. In: Map of the national vegetation of Europe. BfN, Bonn, pp 79–86

Kondracki J, Bohn U (2003) Climatic classification of Europe. In: Map of the national vegetation of Europe. BfN, Bonn, pp 75–78

Kondracki J, Schlüter H (2003) Physicogegraphical classification of Europe. In: Map of the national vegetation of Europe. BfN, Bonn, pp 66–74

Kosmas C, Danalatos NG, López-Bermúdez F, Romero Díaz MA (2002) The effect of land use on soil erosion and land degradation under Mediterranean conditions. In: Geeson NA, Brandt CJ, Thornes JB (eds) Mediterranean desertification: a mosaic of processes and responses. Wiley, Chichester, pp 57–71

Lang G (1994) Quartiäre Vegetationsgeschichte Europa. Methoden und Ergebnisse. G. Fischer, Jena, 462p

Lang G, Rybniček K, Rybničkova E (2003) Late-glacial and Holocene vegetation history of Europe. In: Map of the natural vegetation of Europe. BfN, Bonn, pp 87–104

Meusel H, Jäger E (1992) Vergleichende Chorologie der Zentraleuropäischen Flora. Text Bd. 3. G. Fischer, Jena, 333p

Millána MM, Estrelaa MJ, Sanza MJ, Mantillaa E, Martína M, Pastora F, Salvadora R, Vallejoa R, Alonsob L, Gangoitib G, Ilardiab JL, Navazob M, Albizuric A, Artíñanod B, Cicciolie P, Kallosf G, Carvalhog RA, Andrésh D, Hoffi A, Werhahnj J, Seufertk G, Versinok B (2005) Climatic feedbacks and desertification: the Mediterranean model. J Clim, Boston 18:684–701

Park G (2015) Die Geologie Europas. wbg Academic, Darmstadt, 200p

Pfadenhauer J, Klötzli F (2014) Vegetation der Erde. Grundlagen, Ökologie, Verbreitung. Springer Spektrum, Heidelberg, 645p

Press JR, Short MJ (2016) Flora of Madeira. Pelagic Publishing, London, 574p

Sainz Ollero H, van Staalduinen MA (2012) Iberian Steppes. In: Werger MJA, van Staalduinen MA (eds) Eurasian steppes. Ecological problems and livelihoods in a changing world. Springer, Dordrecht, pp 273–286

Schäfer H (2002) Flora of the Azores. Margraf Verlag, Weikersheim, 264p

Smelansky IE, Tishkov AA (2012) The steppe biome in Russia: ecosystem services, conservation status, and actual challenges. In: Werger MJA, van Staalduinen MA (eds) Eurasian steppes. Ecological problems and livelihoods in a changing world. Springer, Dordrecht, pp 45–102

Sziemer P (2000) Ein kurze Naturgeschichte Madeiras. Ribeiro, Funchal, 288p

Thornes JB (2002) The evolving context of Mediterranean desertification. In: Geeson NA, Brandt CJ, Thornes JB (eds) Mediterranean desertification: a mosaic of processes and responses. Wiley, Chichester, pp 5–12

Walter H, Lieth H (1967) Klimadiagramm – Weltatlas, 3rd edn. G. Fischer, Jena

Walter H, Harnickell E, Müller-Dombois D (1975) Klimadiagramme – Karten der einzelnen Kontinente und ökologische Klimagliederung der Erde. In: Walters H (ed) Vegetationsmonographien der einzelnen Großräume, vol X. G. Fischer, Stuttgart, pp 1–36

Wirthmann A (1970) Zur Klimageomorphologie von Madeira und anderen Atlantikinseln. In: Karlsruher Geographische Hefte, Nr 2., Geographisches Institut der Universität, Karlsruhe

Late Pleistocene and Holocene History of Mammals in Europe

5

Robert S. Sommer

Contents

Abstract

This chapter describes the history of mammal species and their zoogeographical patterns during the Late Pleistocene and Holocene. The scientific results of the past 20 years on biogeography, phylogeny, and population history of mammals in combination with information on the development of climate and environment yield a complex picture of the dynamics of species in the past and offer a better understanding of the Quaternary biogeography of Europe. Overviews of the individual pattern of the different species and the differences among the species are presented. In addition to the extinction pattern of typical Ice Age species like mammoth and cave lion and colonization history of temperate species from glacial refugia like roe deer or beaver, the chapter highlights how successful the different species were during a relatively short period of fundamental climate change. Apart from the extinct species, this is of high relevance for extant species which underwent significant changes of their distribution range during the Pleistocene to Holocene shift like reindeer, saiga antelope, pika, spotted hyena, or arctic fox.

R. S. Sommer (✉)
Department of Landscape Sciences and Geomatics, Neubrandenburg University of Applied Sciences, Rostock, Germany

Department of Zoology, Institute of Biosciences, University of Rostock, Rostock, Germany
e-mail: robert.sommer@uni-rostock.de

K. Hackländer, F. E. Zachos (eds.), *Mammals of Europe - Past, Present, and Future*, Handbook of the Mammals of Europe, https://doi.org/10.1007/978-3-030-00281-7_3

Keywords

Post-glacial · Climate change · Environmental change · Extinction · Recolonization · Glacial refugia · Last Glacial Maximum · Faunal history · Mammals · Palaeoecology · Population history

Introduction

The past 50,000 years of the Late Quaternary (Fig. 1) are characterized by several climatic oscillations and environmental changes (Dansgaard et al. 1993; Huntley et al. 2003; Wohlfarth et al. 2008). The consequences of these environmental dynamics on mammals in Europe over time are documented by a complex spatiotemporal pattern of extirpation, extinction, and (re-)colonization for many species in Europe (Hewitt 2000; Sommer and Nadachowski 2006; Stewart and Cooper 2008; Sommer and Zachos 2009; Stuart and Lister 2012; Crees 2013; Crees et al. 2016; Stuart 2015).

From a biogeographical point of view, the faunal dynamics of mammals in Europe are generally characterized by different main patterns during the Pleistocene/Holocene change:

1. Several typical mammal species of the Ice Age fauna, which were adapted to the steppe-tundra biome, the so-called "mammoth steppe," for example, cave lion *Panthera spelaea* or wholly rhino *Coelodonta antiquitatis*, became extinct in Holarctic regions several 1000 years before or directly at the end of the Pleistocene (Stuart and Lister 2012; Stuart 2015; Cooper et al. 2015).
2. Many extant species, distributed in the Pleistocene landscape and adapted to steppe or tundra environments, such as the arctic fox *Vulpes lagopus*, reindeer *Rangifer tarandus*, saiga *Saiga saiga*, or spotted hyena *Crocuta crocuta*, experienced a local extirpation from Central Europe during the end of the Pleistocene or Early Holocene but were able to maintain their distribution in either recent steppe or tundra biomes in Eurasia or Africa (Nadachowski et al. 2016; Rohland et al. 2005; Sommer et al. 2014).
3. The majority of extant mammal species, adapted to temperate environments (mixed woodlands or open mixed woodlands), such as west-European hedgehog *Erinaceus europaeus*, red deer *Cervus elaphus*, beaver *Castor fiber*, or brown bear *Ursus arctos*, recolonized Central and northern European regions from glacial refugia in southern European regions or adjacent areas (Sommer and Nadachowski 2006; Sommer and Zachos 2009).
4. Species adapted to different habitats like stoat *Mustela erminea* and wolf *Canis lupus* were distributed continuously in Europe throughout the Pleistocene and Holocene. However, these species underwent a considerable population turnover/demographic change triggered by the Last Glacial Maximum or Pleistocene/Holocene change (McDevitt et al. 2012; Pilot et al. 2010; Sommer and Benecke 2005b).

Since the Late Holocene, human influence has had an increasing effect on species, which led to extinction (*Equus ferus*, *Bos primigenius*) or loss of large parts of the natural distribution range in large ungulates (*Alces alces*, *Bison bonasus*) or large carnivores (Crees et al. 2016).

This chapter describes the Late Quaternary history of mammal species in Europe with a focus on species of the extant fauna. Subfossil records of the species from archaeological sites, pattern of DNA or stable isotopes of the species along with records of climate and vegetation dynamics inform our understanding of development of past patterns of mammalian biodiversity in Europe.

Synopsis of Late Quaternary Environmental History

The climate and vegetation history of Europe for the Late Pleistocene and Holocene is relatively well documented (e.g., Dansgaard et al. 1993; Björck et al. 1998; Hubberten et al. 1998; Litt et al. 2001, 2003; Barron et al. 2003; Davis et al.

2003; Huntley and Allen 2003; Renssen et al. 2009). In this chapter, the European environmental history of the last 60,000 years is briefly summarized in order to provide a temporal framework for understanding the colonization history of mammals.

From 60,000 to 27,000 years ago (ka), during the Marine Isotope Stage 3 of the Quaternary, the northern hemisphere was characterized by a generally cold environment punctuated by regular warm intervals, the Greenland Interstadials, which lasted from several 100 years to around 3000 years (Fig. 1). The landscape of Europe north of the glacial refuge areas (Fig. 2b) was characterized by open steppe-tundra (so-called mammoth steppe) with an annual mean temperature of −4 °C to −8 °C (Hubberten et al. 1998). The steppe-tundra contained a mix of plant species of today's steppe as well as tundra biomes and was a unique biome during the Pleistocene with no extant analogue. During warmer Interstadial periods the mean temperature rose abruptly (within a few years) to around 10–16 °C and led to a spread of animal and plant species from the southern refuge areas (e.g., Balkans or Iberia) to at least 50° latitude, significantly changing regional biotic assemblages. During the warmer Interstadial periods (Fig. 1) the European Lowland was characterized by a shrub tundra with occurrence of willow and juniper. South of 50° northern latitude there was an open coniferous park-like landscape.

The Last Glacial Maximum (LGM) from 27 to 18 ka was characterized by the maximum advance of the ice sheets, and all regions of Central Europe were affected by discontinuous permafrost. During this period of cooling, when annual mean temperatures in Greenland were about 21 °C lower than today, cold-adapted species and open landscapes reached their most southerly extent and temperate species became isolated in southern glacial refugia (Sommer and Nadachowski 2006; Sommer et al. 2014), although the degree of range isolation differed depending on their adaptation to certain biomes such as mixed deciduous forests. During the LGM the sea level was about 120 m lower than today because of the large amount of water bound in glaciers (Lambeck et al. 2002).

The onset of the last deglaciation of the northern hemisphere in the Early Late Glacial (18–14.7 ka) began around 18 ka, and by 14 ka most northern parts of what are now Germany and Poland, as well as the Baltic States, were deglaciated. Mean temperatures during the Late Glacial (14.7–11.7 ka) rose by about 12 °C from the beginning of the Greenland Interstadial 1, also known as Bølling/Allerød Interstadial (Fig. 2c), and led to environmental change across the whole northern hemisphere. In Central Europe, the warming induced the expansion of birch *Betula* sp., willow *Salix* sp., and poplar *Populus* sp., and during the Greenland Interstadial 1c–a (Allerød) pine *Pinus* sp. also increased its range

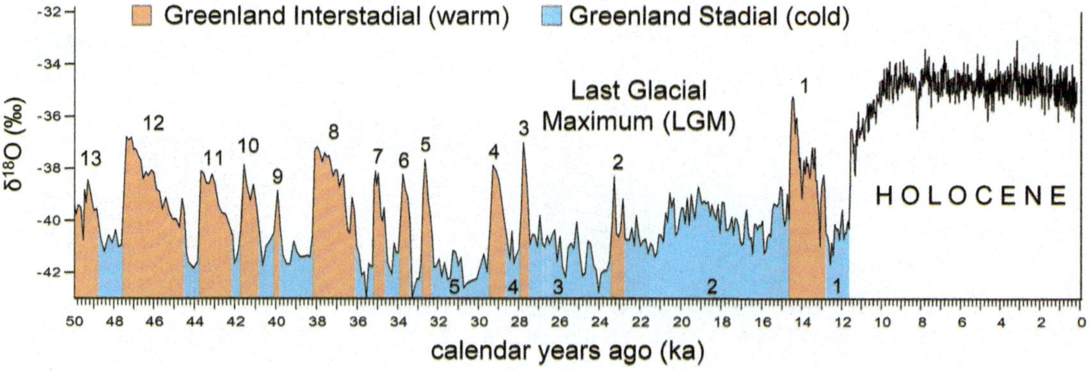

Fig. 1 Climate history of the last 50,000 years based on oxygen isotope ratios $\delta^{18}O$ in ‰ from the GISP2 ice core record (using CalPal, Weninger et al. 2008) and ice core event stratigraphy after Blockley et al. (2012)

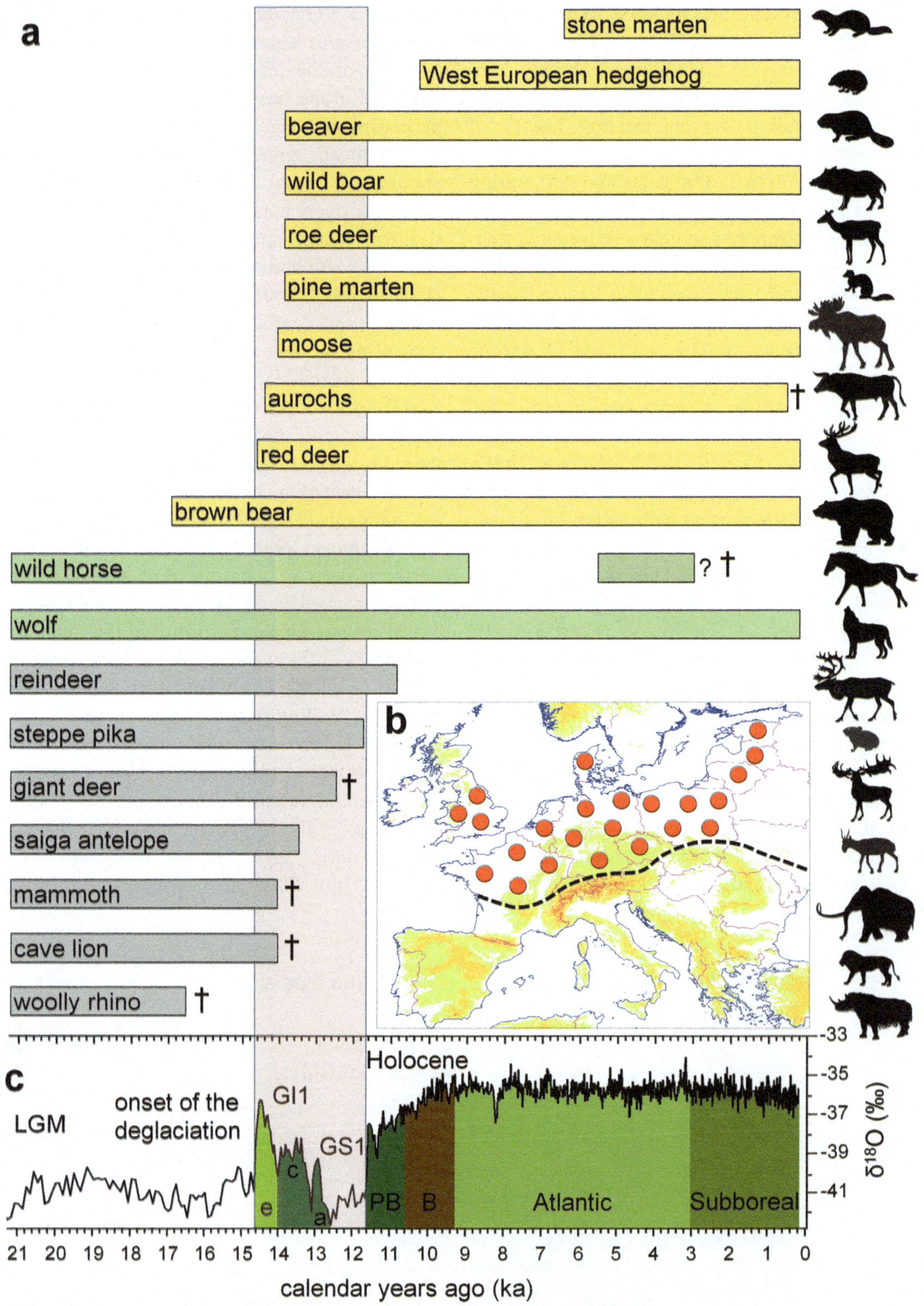

Fig. 2 Chronological comparison of the colonization history of Europe by different mammal species on the basis of the subfossil vertebrate record from archaeological or palaeontological sites since the Last Glacial

(Litt et al. 2001, 2003). However, Europe then experienced a brief cool snap, the Younger Dryas, that lasted for 1000 years and caused the forests which had established during the Greenland Interstadial 1c–a to vanish from northern Central Europe (Theuerkauf and Joosten 2012). The ice core records from Northern Europe (which correspond with temperature changes) unequivocally reflect a rapid rise in temperature at the onset of the Holocene (Fig. 1), which was followed by a slower rise during the Preboreal (PB) and Boreal (Bo) periods (Fig. 2b). This early Holocene warming was associated with a major biome change in Central Europe and the rapid spread of birch and pine, later followed by warm-adapted tree taxa such as hazel *Corylus* sp., oak *Quercus* sp., and elm *Ulmus* sp. During the Holocene Thermal Maximum within the Atlantic climatic period, about 9–5 ka (Fig. 2c and Table 1), when annual mean temperatures were up to 2–3 °C higher than today (Renssen et al. 2009), Central Europe was largely forested, with oak, elm, lime *Tilia* sp., and pine dominating in the lowlands and spruce *Picea* sp., beech *Fagus* sp., and fir *Abies* sp. predominant at higher altitudes. South of 50° northern latitude, open forest steppe communities existed. Since 7 ka ago, the vegetation in Europe has been increasingly influenced by human activities.

Table 1 Event chronology of climatic epochs and environmental change during the Weichselian Glacial and Holocene in Europe

Chronology of Weichselian Glacial and Holocene	Features of environmental change and climate events	Time interval (years ago)
Marine Isotope Stage 4 (MIS 4), Pleniglacial	First maximum cooling of the Weichselian Glacial about 60,000 years ago	75,000–60,000
Isotope Stage 3 (MIS 3), Pleniglacial	Rapid stadial and interstadial oscillations (cf. Fig. 1)	60,000–27,000
Last Glacial Maximum (LGM)	Second maximum cooling of the Weichselian Glacial and maximum and maximal expansion of ice sheets	27,000–18,000
Early Late Glacial	Onset of the last deglaciation	18,000–14,700
Late Glacial Interstadial 1 (Bølling/Allerød warming)	Rapid warming and local return of forests in northern European regions	14,700–12,700
Late Glacial Stadial 1 (younger Dryas)	Cooling oscillation, decline of local woodlands, and subarctic environments in northern Central Europe	12,700–11,700
Early Holocene	Appearance of birch/pine forests, later also hazelnut, in northern Central European regions	11,700–9100
Middle Holocene	Appearance of deciduous woodlands (dominated by oak and lime tree) in northern Central European regions; Holocene Thermal Maximum (HTM) about ca. 9000–4000 years ago; first opening of woodlands by Neolithic settlers about 7000 years ago	9100–5800
Late Holocene	Increasing influence of humans on landscape and environment	5800–present

Fig. 2 (continued) Maximum (LGM). Presence of a species over time is indicated by bars (**a**) from the areas north of the typical refugial regions (**b**) along the climate and environmental history of northern Central European regions. The curve in (**c**) indicates the oxygen isotope ratios $\delta^{18}O$ in ‰ from the GISP2 ice core record (using CalPal, Weninger et al. 2008) and ice core event stratigraphy after Björck et al. (1998). *GI1* Greenland Interstadial 1 (Bølling/Allerød warming epoch), *GS1* Greenland Stadial 1 (Younger Dryas cooling), *PB* Preboreal, *B* Boreal. (**a**) Gray bars: extinct Pleistocene megafauna species or species that underwent substantial range shifts after the Pleistocene; green bars: species that were able to adapt to Pleistocene and Holocene biomes in Europe; yellow bars: temperate species that recolonized Central Europe at the end of the Pleistocene from glacial refugia. The vertical pink bar in (**a, c**) between ca. 14.7 and 11.7 ka ago indicates the time of the Greenland Interstadial 1 warming (GI1) and the Greenland Stadial 1 (GS1). The bars in (**a**) show that the majority of faunal shifts took place during this epoch at the end of the Weichselian Glacial

Pleistocene and Holocene Distribution Dynamics of Mammals in Europe

Ungulates and Other Large Herbivores

With respect to the reconstruction of spatiotemporal dynamics of mammals, ungulates are the most representative species, because they were the main prey of Neanderthals and modern humans in prehistoric times and thus represent the main part of vertebrate bones as food remains of Stone Age hunters at archaeological sites.

The history of the woolly mammoth *Mammuthus primigenius* and the woolly rhino *Coelodonta antiquitatis*, most prominent and dominant members of the Ice Age fauna in Eurasia, has been studied in great detail, most of all in the framework of radiocarbon-based chronologies and population history (Stuart et al. 2004; Stuart and Lister 2012; Palkopoulou et al. 2013; Lister and Sher 2015; Stuart 2015; Kahlke 2015; Chang et al. 2017). The mammoth always occurred in the ice-free parts of Europe during the Weichselian Glacial until 14 ka ago, when it experienced a global collapse caused by environmental change during the Greenland Interstadial 1 and survived only in restricted areas of Northern Siberia and in some parts of the northern Russian Plain until the Early Holocene. On the Wrangel Island in Northeastern Siberia, the mammoth survived until about 4 ka ago (Stuart et al. 2004; Stuart 2015). The endemic European mammoth population of the Weichselian Glacial became extinct after 24 ka ago and was replaced by mammoths of a Siberian genetic clade, which had been colonizing Europe since 34 ka ago (Palkopoulou et al. 2013).

The woolly rhino experienced a local extinction in Europe about 17 ka ago and died out globally about 14 ka ago. Another typical megafauna species of the Ice Age, the giant deer or "Irish elk" *Megaloceros giganteus*, disappeared from Europe during the terminating glacial but persisted until at least 8 ka ago in Western Siberia and European Russia (Stuart et al. 2004; Stuart 2015).

The range of the saiga antelope, an extant steppe dweller of Ponto-Caspian and Asian steppe regions, was restricted to large parts of North Asia and steppe areas north of the Black Sea during the Late Quaternary (Kahlke 2014). It colonized Central and Western Europe only at irregular intervals since the Last Glacial Maximum between 24 and 13 ka ago (Nadachowski et al. 2016; Yalden 1999) and occupied a separate ecological niche compared with other ungulate species of the mammoth steppe (Jürgensen et al. 2017).

The reindeer *Rangifer tarandus*, a key species of the Pleistocene megafauna of the northern hemisphere (Kahlke 2014; Stuart and Lister 2012) and characteristic faunal element of extant European taiga and tundra biomes, was distributed throughout the continent during the Last Glacial, with the exception of the Mediterranean peninsulas, some glaciated parts of Scandinavia, and the northern Caspian region (Kahlke 2014). In contrast to other megafauna species like mammoth or woolly rhino, which died out in most parts of the distribution area after the collapse of the steppe-tundra about 14 ka ago (Stuart and Lister 2012), it showed a much weaker reaction to the loss of the Pleistocene steppe-tundra after the GI1 warming event with a continuous local presence in the northern European Lowlands (Sommer et al. 2014). It survived into the early Holocene until about ca. 11 ka ago and in southern Scandinavia until ca. 10.3 ka ago (Sommer et al. 2014) (Fig. 2a). In southeastern Central Europe, *Rangifer tarandus* quickly became extinct after the LGM, and in Southwestern Europe, its last occurrence was during cold spells of the Younger Dryas period at the end of the Ice Age (Sommer et al. 2014) (Fig. 3b).

Among the extant deer species, the fallow deer *Dama dama* colonized Central Europe only during the Eemian Interglacial (126–115 ka ago). During the Weichselian Glacial it was restricted to Asia Minor, the Balkans (mainly today's Greece), and today's Italy (Yannouli and Trantalidou 1999; Masseti 1996, 1999). Only archaeological sites of Greece and Turkey show a continuous presence from the Late Pleistocene into the Holocene, as displayed in Fig. 3c, which suggests a glacial refugium and thus a natural post-glacial origin of *Dama dama* also in the

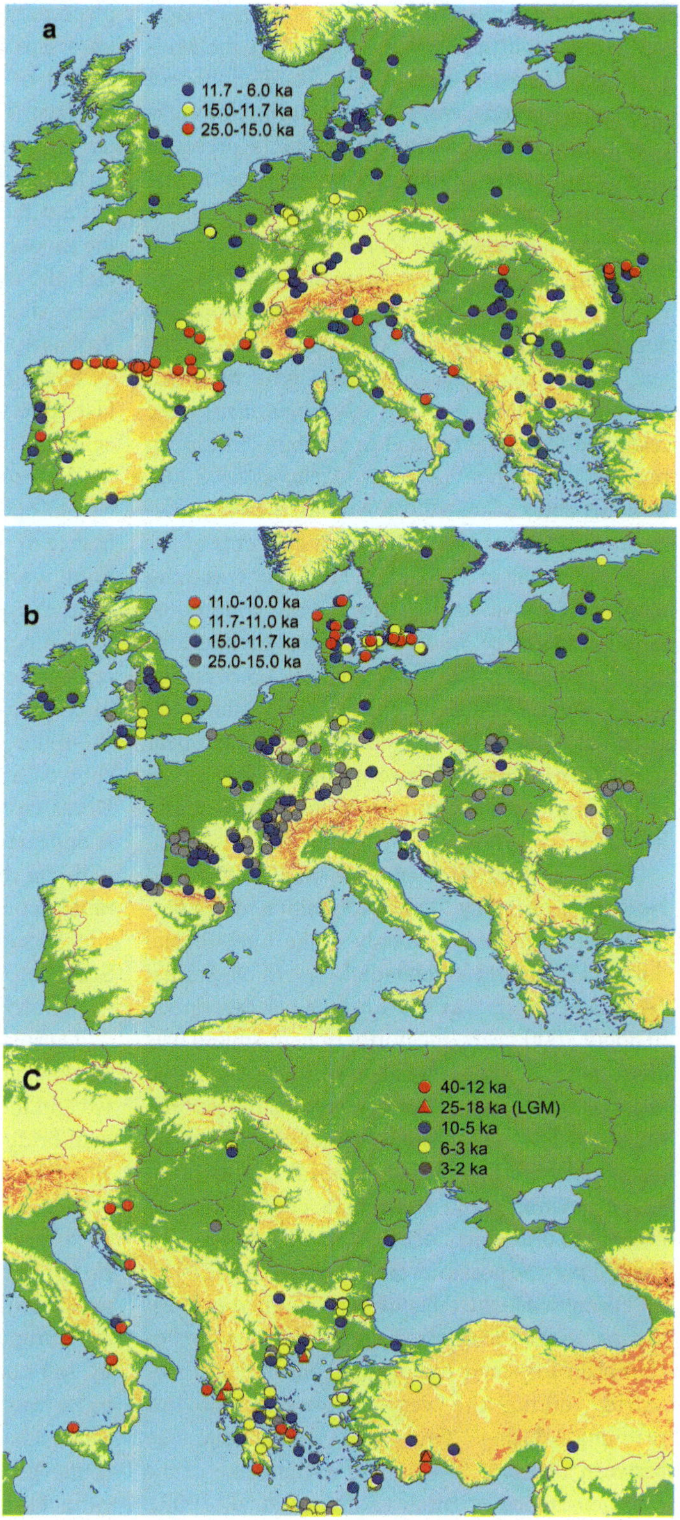

Fig. 3 Pleistocene and Holocene distribution of subfossil remains of different ungulate species in Europe showing different spatiotemporal dynamics and range shifts as a reaction to the Pleistocene/Holocene climate change. (**a**) Roe deer *Capreolus capreolus* as a typical temperate species with a restriction to southern glacial refugia during the LGM, a colonization of southern Central Europe during the Late Glacial, and a Holocene range shift to northern regions. (**b**) Reindeer *Rangifer tarandus* as a species adapted to Pleistocene landscapes that underwent a significant range shift from southern to northern regions at the end of the Pleistocene. (**c**) Fallow deer *Dama dama* as a Ponto-Mediterranean species with glacial refugia in Anatolia and the Balkans. During the Holocene, *Dama dama* was not able to colonize Central Europe out of its glacial refugia. Its present distribution range is largely due to human translocations since Roman Times

Balkans, as discussed in Baker et al. (2017). Bulgaria was colonized by fallow deer since the Mid-Holocene at least 10 ka ago (Fig. 3c). Fallow deer never recolonized Central Europe during the early to Mid-Holocene by natural dispersal. Its presence outside of Ponto-Mediterranean regions is the result of repeated (re-)introductions and translocations by humans since Roman Times (Fig. 3c).

During the Weichselian Glacial temperate species like European roe deer *Capreolus capreolus* (Fig. 3a), red deer *Cervus elaphus*, and moose *Alces alces*, but also wild boar *Sus scrofa* and aurochs *Bos primigenius*, were restricted to glacial refugia in southern European regions (Sommer and Nadachowski 2006; Sommer and Zachos 2009). Among these temperate large mammal species (typical elements of the Holocene fauna of Europe), the red deer was more tolerant of Pleistocene environmental conditions. As a consequence, in warmer Interstadial periods during the Weichselian Glacial, it co-occurred with typical Pleistocene species like mammoth, reindeer, wild horse, spotted hyena, or cave bear in northern regions of Western Europe (Sommer et al. 2008) and Central Europe (Sommer et al. 2008; Musil 2003).

The recolonization process of Central and Northern Europe by the abovementioned temperate artiodactyls began already in the Late Glacial during Greenland Interstadial 1, which started 14.7 ka ago (Fig. 2a). During this warmer epoch of about 2000 years, the northern hemisphere was characterized by fundamental environmental changes which led to the collapse of the steppe-tundra biome and global extinction of woolly rhino and cave lion as well as local extinction of woolly mammoth in Europe (a contribution of humans to the decline of the mammoth in Western Europe may be possible, cf. Lorenzen et al. 2011).

In contrast to the other aforementioned species, the red deer recolonized the western northern European Lowland regions and the British Isles already during the very early stage of this warming during Greenland Interstadial 1e between 14.7 and 14.0 ka ago in the wake of the first occurrence of birch (Sommer et al. 2008). The recolonization of the northern parts of Central Europe by the other species after the LGM took place during the Greenland Interstadial 1a–c (also known as Allerød-warming), in line with a further spread of birch and pine in northern European regions. The moose colonized the British Isles already during the very early Late Glacial stage GI1e (Yalden 1999). While in Central Europe and Britain the red deer shows a similar presence as the moose, i.e., already during the early Late Glacial, the expansion to southern Scandinavia took place only during the Early Holocene (Aaris-Sørensen 2009).

During the Younger Dryas cooling period, the last 1000 years before the beginning of the Holocene, the transformation of the northern regions of the European Lowland from light birch/pine forests to open park-like tundra caused a new dominance of reindeer as well as the last appearance in Southwestern Europe (Sommer et al. 2014) and local extirpation of temperate species.

Temperate ungulate species, for example, represented by the roe deer, survived in light birch/pine forests south of 50° northern latitude (Sommer et al. 2009). The moose, however, showed a more resilient reaction to the Younger Dryas environments in Northern Europe because it appeared in the European Lowlands together with the reindeer, and both species show comparable economic importance in an archaeological site of stone age hunters (Gramsch et al. 2013; Sommer pers. data) and occurred also in Denmark (Schmölcke and Zachos 2005).

During the Early Holocene, also other temperate artiodactyls (re-)colonized southern Scandinavia and the British Isles which were connected by land bridges with the continent since about 10 ka ago. The spatiotemporal complexity of this process is shown, for instance, by red and roe deer (Sommer et al. 2008, 2009). While the aurochs *Bos primigenius* seems to represent a similar temporal recolonization pattern of Central and Northern Europe after the LGM as typically represented by roe deer, the European bison *Bison bonasus* has a more enigmatic faunal history in Europe. Since it has been hypothesized that *Bison bonasus* is the product of a natural hybridization of the extinct steppe bison *Bison priscus* and the aurochs *Bos primigenius* during the Eemian Interglacial c. 120 ka ago (Soubrier et al. 2016), the faunal origin for its Holocene distribution is probably Europe. Although it is recorded since the terminating Late Glacial or Early Holocene in Central and Northern Europe (Benecke 2004, 2005; Aaris-Sørensen 2009), its presence in the vertebrate faunal record from archaeological sites of the early to Mid-Holocene is extremely rare. This, however, might partly be an artifact caused by the fact that fragmented bone remains of aurochs and European bison are difficult to discriminate, and it is likely that an unknown part of bones of the European bison was identified as aurochs in the Holocene Vertebrate faunal record from archaeological sites in Europe (Benecke 2005). Higher frequencies as prey species for humans are recorded only during the Late Holocene in Northeastern Europe, for example, the Baltic States or Belarus (Benecke 2005), which coincides with the core area of the later Holocene distribution area of *Bos bonasus* in eastern Central Europe on the basis of geographic modeling (Kuemmerle et al. 2012).

The biogeographical history of the wild horse *Equus ferus* stands in clear contrast to most other ungulate species in Europe. During the Pleistocene, it was a typical element of the cold-adapted faunal community of the steppe-tundra and experienced a decline in Northern Europe in the course of forest growth and expansion of deciduous forests and a loss of open landscape in the Mid-Holocene (Sommer et al. 2011; Leonardi et al. 2018). In southern European and southern Central European regions, it survived in the open forests and even adapted to open woodlands, as evidenced by the development of a black coat color (Sandoval-Castellanos et al. 2017). After the opening of the primeval forests by Neolithic farmers, *Equus ferus* could recolonize the European Lowland to a certain extent and was replaced by the domestic horse from the Bronze Age onward (Sommer et al. 2011, 2018).

Carnivorans

Among the carnivorans, very prominent members of the Ice Age megafauna were distributed in Europe. The spotted hyena *Crocuta crocuta* was distributed in Pleistocene faunal communities in Eurasia. The traditional view that it was a separate Pleistocene hyena on species or subspecies level was rejected by Rohland et al. (2005), who show that cave hyenas are a genetic clade of African spotted hyenas, arisen through dispersal and separation in Eurasian environments during the Late Pleistocene. Interestingly, it represents the earliest extirpation event among the megafauna carnivorans during the Last Glacial in Eurasia; the latest record is dated about 30 ka ago in Italy (Stuart and Lister 2014; Stuart 2015).

The cave bear *Ursus spelaeus*, well known from numerous Pleistocene fossil bone assemblages in European cave systems (Münzel et al. 2011), became extinct in Europe (and globally) about 28 ka ago (Pacher and Stuart 2009). Interestingly, long before this event, a different species of cave bear, *Ursus ingressus*, hypothesized to be a separate cave bear species in Eastern Europe only on the basis of ancient DNA (Knapp et al. 2009), appeared in the Alpine region and coexisted with the cave bear at least for 4500 years (Münzel et al. 2011). *Ursus ingressus* replaced *Ursus spelaeus* in the Alpine region about 30 ka ago but outlived *Ursus spelaeus* only by about 2000 years (Münzel et al. 2011). Genetic data from cave bears in Northern Spain (Fortes et al. 2016) show that each cave was colonized by a unique genetic lineage of *Ursus spelaeus* suggesting an extreme fidelity of cave bears to their birth sites (homing behavior). Brown bears *Ursus arctos*, which also occurred in the same region during the Weichselian Glacial and the LGM (Sommer and Benecke 2005a), do not show this strong association of genetic lineage and cave locality; thus, in times of increasing competition between humans and brown bears, it is very likely that this could have contributed to cave bear extinction (Fortes et al. 2016). The brown bear was the earliest colonizer of the regions north of potential glacial refugia and reached today's British Isles during the Early Late Glacial following the onset of the deglaciation (Sommer and Benecke 2005a; Fig. 2a). *Ursus arctos* colonized Northwestern Europe sporadically for probably shorter periods during warmer interstadials, several 1000 years before the LGM (Sommer and Benecke 2005a). The polar bear

Ursus maritimus was distributed to an unknown extent in the periglacial landscape of the Weichselian Glacial, probably where northern ice shields met open waters of the Atlantic. The youngest record is from the Late Glacial in Denmark about 12.4–12.9 ka ago (Aaris-Sørensen 2009).

Among the mustelids, the stoat *Mustela erminea* and the least weasel *M. nivalis* were distributed in both full glacial and forested Holocene landscapes of Europe (Sommer and Benecke 2004). The temperate mustelid species, isolated into glacial refugia in Southern Europe during the Weichselian Glacial like the forest-dependent pine marten *Martes martes*, colonized Central Europe and southern Scandinavia during the Allerød-warming in the Late Glacial when these regions were increasingly covered by open birch-pine woods (Sommer and Benecke 2004). However, it is unclear to what extent these species underwent local extinction during the unfavorable environmental conditions of the Younger Dryas.

The biogeographical pattern of the stone marten *Martes foina* is in strong contrast to *M. martes*, because the colonization of the stone marten during the Mid-Holocene followed the spread of agriculture and farming with an assumed origin in Asia Minor; thus, the adaptation of *M. foina* to human settlements has its origin already in the Neolithic (Sommer and Benecke 2004). In contrast to most other carnivoran species the Eurasian otter *Lutra lutra* only recolonized Central Europe during the Holocene about 10 ka ago (Sommer and Benecke 2004; Aaris-Sørensen 2009), perhaps because it was only present in a single glacial refuge in the Apennine peninsula. The Alps would have been a migration barrier during this recolonization process (Sommer and Benecke 2004). The relatively low genetic diversity of otter populations in Europe (Mucci et al. 2010; Honnen et al. 2011) may be a consequence of this. The Pleistocene and Holocene distribution dynamics of canids are described by Sommer and Benecke (2005b).

The cave lion *Panthera spelaea*, which was a classical faunal element and very widespread in Eurasia during the Late Pleistocene (Stuart and Lister 2011), became extinct about 14 ka ago (like the woolly rhino), caused by the collapse of the steppe-tundra, their typical biome (Stuart and Lister 2011, 2012). After global extinction of the cave lion, the modern lion (*Panthera leo*) reached Eastern Europe about 8000 years ago and colonized today's Greece, Bulgaria, and the Pannonian Basin during the Mid-Holocene (Sommer and Benecke 2006; Masseti and Mazza 2013). The latest presence of *Panthera leo* in Europe in Bulgaria and Greece is dated to the Bronze and Iron Ages (Ninov 1999; Sommer and Benecke 2006). The Eurasian Lynx *Lynx lynx* was climatically less specialized and was regularly present in Central Europe in the Weichselian Pleniglacial before the LGM (Sommer and Benecke 2006). During the Late Glacial it was one of the earliest carnivorans to colonize the northern regions, and its distribution included Northern Iberia where it occurred sympatrically with the Iberian lynx *Lynx pardinus* during this period. From a zoogeographical point of view, the Iberian lynx shows an interesting palaeoendemism, because during the LGM it was distributed also in Northern Italy together with *L. lynx* (Rodríguez-Varela et al. 2015), and its Holocene range included parts of Southern France until about 2000 years ago (Sommer and Benecke 2006; Rodríguez-Varela et al. 2015). Since the Late Holocene, the range of *L. pardinus* has been restricted to Iberia (Sommer and Benecke 2006). The wildcat *Felis silvestris* reached Central Europe out of southern refugia during the Late Glacial Allerød-warming in the wake of the returning forests. It is unknown to what extent the species survived the Younger Dryas cooling period north of the Alps (Sommer and Benecke 2006). During the Holocene Thermal Maximum (HTM) the wildcat was also distributed in Scandinavia but vanished again due to the cold climate after the HTM (Sommer and Benecke 2006).

Rodents and Small Mammals

In contrast to carnivorans and ungulates, the biogeographic pattern of Pleistocene and Holocene distribution dynamics of smaller mammals like rodents or bats in Europe has been studied much less. One very good scientific reference is the unpublished PhD thesis of J. Fahlke (2009). The occurrence of subfossil bones of small mammals in Pleistocene and Holocene layers often does not indicate a clear climatic preference, which is also caused by the fact that small and lightweight mammal bones may be frequently relocated in sediment layers over time (von Koenigswald 1974, 1977, 1984a, b). However, when assessing (and re-evaluating) large numbers of subfossil records of a species, the spatiotemporal dynamics in reaction to past climate change can be understood much better (Sommer 2007).

Among the Pleistocene small mammal fauna of Central Europe, the Norwegian lemming *Lemmus lemmus* and the collared lemmings *Dicrostonyx* spp. were typical cold-adapted species which disappeared from Central Europe and survived in recent northern Eurasian boreal and tundra regions. However, there is no detailed information about the exact time of local extinction (whether at the end of the Pleistocene or the beginning of the Holocene). Prost et al. (2010a) showed that previous climate warming events had a strong influence on genetic diversity and population size of collared lemmings *D. torquatus* in arctic ecosystems. An interesting zoogeographical case is the narrow-headed vole *Microtus gregalis*. Presently distributed in both tundra and steppe environments of Asia, it was considered to be an indicator of cold climate in Central European faunal history (von Koenigswald 2002) and already reached northern European regions during the Weichselian Glacial (Fahlke 2009). One of the most prominent representatives of past Pleistocene faunal communities in Europe is the steppe pika *Ochotona pusilla*. As a steppe dweller, it is presently restricted to steppe regions of Kazakhstan and Russia but colonized large parts of the European steppe-tundra, including parts of Northern Europe (Fahlke 2009). *Ochotona pusilla* was a frequent small mammal species in colder stages of the Pleistocene in Europe and disappeared from today's Germany and the British Isles at the end of the Younger Dryas Stadial (Street and Baales 1999; Fahlke 2009; Yalden 1999). Whereas the steppe pika underwent a postglacial contraction of its range, the mountain hare *Lepus timidus*, which was continuously distributed throughout Central Europe during the Weichselian Glacial, experienced a separation which resulted in a disjunct arctic-alpine distribution (plus the British Isles). Assuming that the discrimination of numerous Pleistocene and Holocene subfossil bone remains of hares is correct, they reveal a clear pattern of an exclusive presence of mountain hares in Pleniglacial and Late Glacial epochs in Central Europe and southern Scandinavia. During the Early Holocene, it was then replaced by the brown hare *Lepus europaeus*. Southern Scandinavia was first colonized during the Neolithic, possibly in the wake of the opening of woodlands by Stone Age farmers.

The common hamster was part of both Pleistocene and Holocene faunal communities in Europe (Fahlke 2009). As a steppe dweller, it may have benefited, similarly to the wild horse, from opening of woodlands by Neolithic farmers since the end of the Mid-Holocene.

It is interesting that the Russian Desman *Desmana moschata*, presently restricted to semiaquatic habitats mainly at the riverbanks of the Volga, Don, and Ural in Russia, colonized the northern European Lowland and southern Scandinavia during the whole Late Glacial period and disappeared with the beginning of the Holocene. In contrast to that, the European Mole *Talpa europaea* was present during Pleni- and Late Glacial stages of the Pleistocene north of the Alps (Fahlke 2009) but is first recorded from the European Lowland and Scandinavia during the Holocene (Fahlke 2009; Aaris-Sørensen 2009).

Among the temperate species, the beaver *Castor fiber* shows an interesting recolonization pattern very similar to temperate ungulate species like moose, red deer, and roe deer. At first, a post-LGM recolonization of parts of the European Lowlands (including southern Scandinavia) during the Greenland Interstadial 1 took place following the sudden climate warming and increase in forests and appearance of typical food plants like birch and willow. Moreover, subfossil wood with visible cut marks from beavers from Northeastern Germany has been dated to the Greenland Interstadial 1c–a (Lampe et al. 2016). During the

Younger Dryas cooling period there is a gap in the beaver subfossil record in northern regions of Central Europe (Fahlke 2009; Aaris-Sørensen 2009) which may indicate local extinction caused by unfavorable climatic conditions. However, since the early Holocene the beaver has continuously been documented in the subfossil vertebrate record of northern European regions (Benecke 1999; Aaris-Sørensen 2009).

The bank vole *Myodes glareolus* is known as a typical Holocene small mammal species in Europe. Today the species is mainly associated with woodlands. The species shows a remarkable flexibility in climatic preferences in its post-LGM distribution history and colonized the lower mountain ranges of today's Germany from the Early Late Glacial onward (Fahlke 2009). The phylogeographic pattern clearly suggests a post-LGM colonization out of a Carpathian refuge (Kotlík et al. 2006). The survival of the LGM in the Carpathian region was also discovered in brown bear (Sommer and Benecke 2005a) and numerous temperate mammals such as red fox, pine marten, red deer, roe deer, and moose (Sommer and Nadachowski 2006). The strongly forest-dependent edible dormouse *Glis glis* was restricted to deciduous forest refugia in the Mediterranean/Pontomediterranean regions of Southern Europe during the LGM and first colonized the area north of the Alps during the Boreal period in the wake of the northward spread of deciduous forests (Hürner et al. 2010). The Eurasian red squirrel *Sciurus vulgaris*, ecologically more flexible than the edible dormouse due to its adaptation to mixed wood and taiga biomes, was also an exclusive Holocene colonizer of Central Europe (Fahlke 2009), and it can be assumed that the colonization process took place very rapidly during the Preboreal period. However, the species is underrepresented in the vertebrate faunal record and thus it is questionable if the first record in southern Scandinavia with an age of 9.5 ka ago really represents its first appearance in Northern Europe.

The west-European hedgehog *Erinaceus europaeus* is an example of a Holocene immigrant (Fig. 2a) that colonized Central Europe and southern Scandinavia first during the Early Holocene about 11–10 ka ago and shows (in contrast to other smaller mammal species) a clear preference of warmer Interglacial climate (Sommer 2007). The white-breasted hedgehog *Erinaceus roumanicus*, which recolonized eastern parts of Central Europe out of a glacial refuge from the Balkans, is only rarely found in the vertebrate record, but a similar temporal pattern as in *E. europaeus* has been suggested (Sommer 2007).

The common shrew *Sorex araneus* as well as the Eurasian Pygmy Shrew *Sorex minutus* show no strong climatic preference and appear frequently in Pleistocene as well as Holocene faunal communities. The common shrew developed a large ecomorph in Alpine regions during the Late Glacial which led to the assumption of a separate species *Sorex macrognathus*. However, the populations of this large morph of *S. araneus* were replaced by other populations after the Pleistocene/Holocene change (Prost et al. 2013). This example demonstrates the high morphological plasticity and the dynamics of population turnover in a small mammal species during climate change at the end of the Pleistocene. A similar case is the suggested extinction of the Don-hare *Lepus tanaiticus* at the end of the Pleistocene in the Russian Plains, which has genetically been identified as a morphotype of *L. timidus* (Prost et al. 2010b). In light of Prost et al., it would be interesting to check if the large Pleistocene hamster species *Cricetus major*, extinct in Europe by the beginning of the Greenland Interstadial 1 (Fahlke 2009), may rather be an ecomorph of *Cricetus cricetus* than a separate hamster species.

Concluding Remarks

Increasingly, research activities and the development of new methods pertaining to ancient DNA over the last decades have led to a "revolution" of our understanding of species dynamics and faunal development, yielding new results on species origins, dynamics of species at the population level, morphology, and even behavior of species.

The Pleistocene and Holocene history of mammals in Europe in light of modern research increasingly suggests individualistic dynamics of species in response to the climatic and environmental change at the end of the Pleistocene or the Pleistocene/Holocene shift.

Without doubt, the Last Glacial Maximum with its impact on phylogeography and as a driver of speciation processes shaped the roots of our extant European fauna. Apart from the Last Glacial Maximum and the beginning of the Holocene, the Greenland Interstadial 1 (Bølling/Allerød warming) was the most important climatic event for the timing of the faunal shifts about 10,000 years after the culmination point of the Last Glacial Maximum. Fundamental changes in the faunal history of extant temperate European mammal species and basic features of the extant distribution patterns of these species (like the colonization of Central European regions north of the Alps from glacial refugia) took place during the Bølling/Allerød warming. The intensive and detailed analysis of subfossil records of mammals, mainly from archaeological and palaeontological sites, indicates that only few species like hedgehogs, otter, or the forest-dependent edible dormouse are "real" Holocene colonizers. The majority of species, for example, most of the carnivorans and ungulates, recolonized important Central European regions already during the Greenland Interstadial 1 (Fig. 2), about 2000 years before the start of the Holocene. Thus, the impact of this climatic event, which caused the local extinction of the mammoth in Western Europe, the recolonization of Central Europe by temperate species and the global extinction of the cave lion and the woolly rhino during the Late Glacial, warrants further research efforts in Quaternary zoogeography and palaeoecology. Together with the dynamics of vegetation and Stone Age humans this will yield a much more detailed picture of the multifaceted connections and interactions of mammals, climate, landscape change, and humans in the past.

The striking combination of the spatiotemporal distribution pattern on the basis of subfossil records (at species level) and information from ancient DNA (at population or species level) has led, and is still leading, to a much better understanding of European biogeography, and it may also be key to our understanding of how the reconstruction of past biogeography might inform us on future developments in the face of global change.

Acknowledgments I am grateful to Frank Zachos for helpful suggestions and comments on an earlier version of the text.

References

Aaris-Sørensen K (2009) Diversity and dynamics of the mammalian fauna in Denmark throughout the last glacial–interglacial cycle, 115–0 kyr BP. Fossils Strata 57:1–59

Baker KH, Gray HWI, Ramovs V, Mertzanidou D, Akın Pekşen Ç, Bilgin CC, Sykes N, Hoelzel AR (2017) Strong population structure in a species manipulated by humans since the Neolithic: the European fallow deer (*Dama dama dama*). Heredity 119:16–26

Barron E, van Andel TH, Pollard D (2003) Glacial environments II. Reconstructing the climate of Europe in the last glaciation. In: van Andel TH, Davies W (eds) Neanderthals and modern humans in the European landscape during the last glaciation: archaeological results of the Stage 3 Project. McDonald Institute Monographs, Cambridge

Benecke N (1999) Die Jungpleistozäne und holozäne Tierwelt Mecklenburg-Vorpommerns – Faunenhistorische und kulturgeschichtliche Befunde. Documenta Naturae 124:1–198

Benecke N (2004) Faunal succession in the lowlands of northern Central Europe at the Pleistocene–Holocene transition. In: Terberger T, Eriksen BV (eds) Hunters in a changing world. Environment and archaeology of the Pleistocene–Holocene transition (ca. 11.000–9000 BC) in northern Central Europe. Marie Leidorf Publisher, Rahden

Benecke N (2005) The Holocene distribution of European bison: the archaeozoological record. Munibe 57:421–428

Björck S, Walker MJC, Cwynar LC, Johnsen S, Knudsen K-L, Lowe JJ, Wohlfarth B (1998) An event stratigraphy for the Last Termination in the North Atlantic region based on the Greenland ice core record: a proposal by the INTIMATE group. J Quat Sci 13:283–292

Blockley SPE, Lane CS, Hardiman M, Rasmussen SO, Seierstad IK, Steffensen JP, Svensson A, Lotter AF, Turney C, Ramsey C, INTIMATE Members (2012) Synchronisation of palaeoenvironmental records over the last 60,000 years, and an extended INTIMATE event stratigraphy to 48,000 b2k. Quat Sci Rev 36:2–10

Chang D, Knapp M, Enk J, Lippold S, Kircher M, Lister A, MacPhee RDE, Widga C, Czechowski P, Sommer RS, Hodges E, Stümpel N, Barnes I, Dalén L, Derevianko A, Germonpré M, Hillebrand-Voiculescu A, Constantin S, Kuznetsova T, Mol D, Rathgeber T, Rosendahl W,

Tikhonov AN, Willerslev E, Hannon G, Lalueza-Fox C, Joger U, Poinar H, Hofreiter M, Shapiro B (2017) The evolutionary and phylogeographic history of woolly mammoths: a comprehensive mitogenomic analysis. Sci Rep 7:44585

Cooper A, Turney C, Hughen KA, Brook BW, McDonald HG, Bradshaw CJA (2015) Abrupt warming events drove Late Pleistocene Holarctic megafaunal turnover. Science 349:602–606

Crees JJ (2013) Dynamics of large mammal range shifts and extinction: evidence from the Holocene of Europe. PhD thesis, Imperial College London and Institute of Zoology, ZSL

Crees JJ, Carbon C, Sommer RS, Benecke N, Turvey S (2016) Millennial-scale faunal record reveals differential resilience of European large mammals to human impacts across the Holocene. Proc R Soc B 283:1–9

Dansgaard W, Johnsen SJ, Clausen HB, Dahl-Jensen D, Gundestrup NS, Hammer CU, Hvidberg CS, Steffensen JP, Sveinbjörnsdottir AE, Jouzel J, Bond G (1993) Evidence for general instability of past climate from a 250-kyr ice-core record. Nature 364:218–220

Davis BAS, Brewer S, Stevenson AC, Guiot J (2003) The temperature of Europe during the Holocene reconstructed from pollen data. Quat Sci Rev 22:1701–1716

Fahlke JM (2009) Der Austausch der terrestrischen Säugetierfauna an der Pleistozän/Holozän-Grenze in Mitteleuropa. Unpublished PhD thesis, University of Bonn, pp 279

Fortes GG, Grandal-d'Anglade A, Kolbe B, Fernandes D, Meleg IN, GarciaVazquez A, Pinto-Llona AC, Constantin S, de Torres TJ, Ortiz JE, Frischauf C, Rabeder G, Hofreiter M, Barlow A (2016) Ancient DNA reveals differences in behaviour and sociality between brown bears and extinct cave bears. Mol Ecol 25:4907–4918

Gramsch B, Beran J, Hanik S, Sommer RS (2013) A Palaeolithic fishhook made of ivory and the earliest fishhook tradition in Europe. J Archaeol Sci 40:2458–2463

Hewitt GM (2000) The genetic legacy of the Quaternary ice ages. Nature 405:907–913

Honnen A-C, Petersen B, Kaßler L, Elmeros M, Roos A, Sommer RS, Zachos FE (2011) Genetic structure of Eurasian otter (*Lutra lutra*, Carnivora: Mustelidae) populations from the western Baltic Sea region and its implications for the recolonisation of Northwestern Germany. J Zool Syst Evol Res 48:169–175

Hubberten HW, Andrev A, Astakhov VI, Demikov I, Dowdeswell JA, Henriksen M, Hjort C, Houmark-Nielsen M, Jakobsson M, Kuzmina S, Larsen E, Lunkka JP, Lysa A, Mangerud J, Moller P, Saarnisto M, Schirrmeister L, Sher AV, Siegert C, Siegert MJ, Huijzer B, Vandenberghe J (1998) Climatic reconstruction of the Weichselian Pleniglacial in Northwestern and Central Europe. J Quat Sci 13:391–417

Huntley B, Allen JRM (2003) Glacial environments III: palaeo-vegetation patterns in Late Glacial Europe. In: van Andel TH, Davies W (eds) Neanderthals and modern humans in the European landscape during the last glaciation: archaeological results of the Stage 3 Project. McDonald Institute Monographs, Cambridge

Huntley B, Alfano MJ, Allen JRM, Pollard D, Tzedakis PC, de Beaulieu JL, Grüger E, Watts B (2003) European vegetation during Marine Oxygen Isotope Stage-3. Quat Res 59:159–212

Hürner H, Krystufek B, Sara M, Ribas A, Ruch T, Sommer RS, Ivashkina V, Michaux J (2010) Mitochondrial phylogeography of the edible dormouse (*Glis glis*) in the western Palearctic region. J Mammal 91:233–242

Jürgensen J, Drucker DG, Stuart AJ, Schneider M, Buuveibaatar B, Bocherens H (2017) Diet and habitat of the saiga antelope during the late Quaternary using stable carbon and nitrogen isotope ratios. Quat Sci Rev 160:150–161

Kahlke RD (2014) The origin of Eurasian mammoth faunas (Mammuthus–Coelodonta Faunal Complex). Quat Sci Rev 96:32–49

Kahlke RD (2015) The maximum geographic extension of Late Pleistocene *Mammuthus primigenius* (Proboscidea, Mammalia) and its limiting factors. Quat Int 379:147–154

Knapp M, Rohland N, Weinstock J, Baryshnikov G, Sher A, Nagel D, Rabeder G, Pinhasi R, Schmidt HA, Hofreiter M (2009) First DNA sequences from Asian cave bear fossils reveal deep divergences and complex phylogeographic patterns. Mol Ecol 18:1225–1238

Kotlík P, Deffontaine V, Mascheretti S, Zima J, Michaux JR, Searle JB (2006) A northern glacial refugium for bank voles (*Clethrionomys glareolus*). Proc Natl Acad Sci U S A 103:14860–14864

Kuemmerle T, Hickler T, Olofsson J, Schurgers G, Radeloff VC (2012) Reconstructing range dynamics and range fragmentation of European bison for the last 8000 years. Divers Distrib 18:47–59

Lambeck K, Esat MT, Potter E-K (2002) Links between climate and sea levels for the past three million years. Nature 419:199–206

Lampe R, Jahnke W, Schuldt M, Meng S, Lampe M (2016) Multiproxy-Untersuchungen zur Paläoökologie und -hydrologie ines spätglazial- bis frühholozänen Flachsees im nordostdeutschen Küstengebiet (Glowe-Paläosee/Insel Rügen). E&G Quat Sci J 65:41–75

Leonardi M, Boschin F, Giampoudakis K, Beyer RM, Krapp M, Bendrey R, Sommer RS, Boscato P, Manica A, Nogues-Bravo D, Orlando L (2018) Late Quaternary horses in Eurasia in the face of climate and vegetation change. Sci Adv 4:1–11

Lister AM, Sher AV (2015) Evolution and dispersal of mammoths across the Northern Hemisphere. Sci 350 (6262):805–809

Litt T, Brauer A, Goslar T, Merkt J, Balaga K, Müller H, Ralska-Jasiewiczowa M, Stebich M, Negendank JFW (2001) Correlation and synchronisation of Lateglacial continental sequences in northern Central Europe based on annually laminated lacustrine sediments. Quat Sci Rev 20:1233–1249

Litt T, Schminke HU, Kromer B (2003) Environmental response to climatic and volcanic events in Central Europe during the Weichselian Lateglacial. Quat Sci Rev 22:7–32

Lorenzen ED, Nogués-Bravo D, Orlando L, Weinstock J, Binladen J, Marske KA, Haile J, Borregaard MK, Ugan A, Yoccoz NG, Coissac E, Gussarova G, Brochmann C, Taberlet P, Gilbert MTP, Nielsen R, Ho SYW, Goebel T, Graf KE, Byers D, Stenderup JT, Rasmussen M, Campos P, Leonard J, Koepfli K-P, Froese D, Zazula G, Stafford T, Higham T, Munch K, Aaris-Sørensen K, Baryshnikov G, Batra P, Burns JA, Conard N, Davydov SP, van Geel B, Gravendeel B, Hockett B, Jenkins D, Kosintsev P, Kuznetsova T, Xulong L, Martin L, Mol D, Meldgaard M, Stephan E, Sablin M, Sommer RS, Sipko T, Scott E, Tikhonov A, Telka A, Willerslev R, Wayne RK, Cooper A, Hofreiter M, Sher A, Shapiro B, Rahbek C, Willerslev E (2011) Species-specific responses of Late Quaternary megafauna to climate and humans. Nature 479:359–364

Masseti M (1996) The postglacial diffusion of the genus *Dama* Frisch, 1775, in the Mediterranean region. Suppl Ric Biol Selvaggina 25:7–29

Masseti M (1999) The fallow deer, *Dama dama* L., 1758, in the Aegean region. Contrib Zoogeogr Ecol East Mediterr Reg 1(Suppl):17–30

Masseti M, Mazza PPA (2013) Western European Quaternary lions: new working hypotheses. Biol J Linn Soc 109:66–77

McDevitt AD, Zub K, Kawałko A, Oliver MK, Herman JS, Wójcik JM (2012) Climate and refugial origin influence the mitochondrial lineage distribution of weasels (*Mustela nivalis*) in a phylogeographic suture zone. Biol J Linn Soc 106:57–69

Mucci N, Arrendal J, Ansorge H, Bailey M, Bodner M, Delibes M, Ferrando A, Fournier P, Fournier C, Godoy J, Hajkova P, Hauer S, Heggberget TM, Heidecke D, Kirjavainen H, Krueger H-H, Kvaloy K, Lafontaine L, Lanszki J, Lemarchand C, Liukko U-M, Loeschcke V, Ludwig G, Madsen AB, Mercier L, Ozolins J, Paunovic M, Pertoldi C, Piriz A, Prigioni C, Santos-Reis M, Luis TS, Stjernberg T, Schmid H, Suchentrunk F, Teubner J, Tornberg R, Zinke O, Randi E (2010) Genetic diversity and landscape genetic structure of otter (*Lutra lutra*) populations in Europe. Conserv Genet 11:583–599

Münzel SC, Stiller M, Hofreiter M, Mittnik A, Conard NJ, Bocherens H (2011) Pleistocene bears in the Swabian Jura (Germany): genetic replacement, ecological displacement, extinctions and survival. Quat Int 245:225–237

Musil R (2003) The middle and upper Palaeolithic game suite in Central and Southeastern Europe. In: van Andel TH, Davies W (eds) Neanderthals and modern humans in the European landscape during the Last Glaciation: archaeological results of the Stage 3 Project. McDonald Institute Monographs, Cambridge

Nadachowski A, Lipecki G, Ratajczak U, Stefaniak K, Wojtal P (2016) Dispersal events of the saiga antelope (*Saiga tatarica*) in Central Europe in response to the climatic fluctuations in MIS 2 and the early part of MIS 1. Quat Int 420:357–362

Ninov LK (1999) Vergleichende Untersuchungen zur Jagd und zum Jagdwild während des Neolithikums und Äneolithikums in Bulgarien. Archäologie in Eurasien 6:323–338

Pacher M, Stuart AJ (2009) Extinction chronology and palaeobiology of the cave bear (*Ursus spelaeus*). Boreas 38:189–206

Palkopoulou E, Dalen L, Lister AM, Vartanyan S, Sablin M, Sher A, Nyström Edmark V, Brandström MD, Germonpre M, Barnes I, Thomas JA (2013) Holarctic genetic structure and range dynamics in the woolly mammoth. Proc R Soc B 280:20131910

Pilot M, Branicki W, Jędrzejewski W, Goszczyński J, Jędrzejewska B, Dykyy I, Shkvyrya M, Tsingarska E (2010) Phylogeographic history of grey wolves in Europe. BMC Evol Biol 10:104

Prost S, Smirnov N, Federov VB, Sommer RS, Stiller M, Nagel D, Knapp M, Hofreiter M (2010a) Influence of climate warming on Arctic mammals? New insights from ancient DNA studies of the collared lemming *Dicrostonyx torquatus*. PLoS One 5:1–11

Prost S, Knapp M, Flemming J, Hufthammer AK, Kosintsev P, Stiller M, Hofreiter M (2010b) A phantom extinction? New insights into extinction dynamics of the Don-hare *Lepus tanaiticus*. J Evol Biol 23:2022–2029

Prost S, Klietmann J, van Kolfschoten T, Vrieling K, Stiller M, Nagel D, Rabeder G, Guralnick R, Waltari E, Hofreiter M, Sommer RS (2013) Effects of Late Quaternary climate change on Palearctic shrews. Glob Chang Biol 19:1865–1874

Renssen H, Seppä H, Heiri O, Roche DM, Goosse H, Fichtfet T (2009) The spatial and temporal complexity of the Holocene thermal maximum. Nat Geosci 2:411–414

Rodríguez-Varela R, Tagliacozzo A, Ureña I, García N, Crégut-Bonnoure E, Mannino MA, Arsuaga JL, Valdiosera C (2015) Ancient DNA evidence of Iberian lynx palaeoendemism. Quat Sci Rev 112:172–180

Rohland N, Pollack JL, Nagel D, Beauval C, Airvaux J, Pääbo S, Hofreiter M (2005) The population history of extant and extinct hyenas. Mol Biol Evol 22:2435–2443

Sandoval-Castellanos E, Wutke S, Gonzalez-Salazar C, Ludwig A (2017) Coat colour adaptation of post-glacial horses to increasing forest vegetation. Nat Ecol Evol 1:1816–1819

Schmölcke U, Zachos FE (2005) Holocene distribution and extinction of the moose (*Alces alces*, Cervidae) in Central Europe. Mamm Biol 70:329–344

Sommer RS (2007) When east met west: the sub-fossil footprints of the west European hedgehog and the northern white-breasted hedgehog during the Late Quaternary in Europe. J Zool 271:82–89

Sommer RS, Benecke N (2004) Late- and post-glacial history of the Mustelidae in Europe (Mustelidae). Mammal Rev 34:249–284

Sommer RS, Benecke N (2005a) The recolonisation of Europe by brown bears *Ursus arctos* Linnaeus, 1758 after the Last Glacial Maximum. Mammal Rev 35:156–164
Sommer RS, Benecke N (2005b) Late-Pleistocene and early Holocene history of the canid fauna of Europe (Canidae). Mamm Biol 70:227–241
Sommer RS, Benecke N (2006) Late-glacial and early Holocene colonisation history of felids in Europe (Felidae): a review. J Zool 269:7–20
Sommer RS, Nadachowski A (2006) Glacial refugia of mammals in Europe: evidence from fossil records. Mammal Rev 36:251–266
Sommer RS, Zachos FE (2009) Fossil evidence and phylogeography of temperate species: "glacial refugia" and postglacial recolonization. J Biogeogr 36:2013–2020
Sommer RS, Zachos FE, Street M, Jöris O, Skog A, Benecke N (2008) Late Quaternary distribution dynamics and phylogeography of the red deer (*Cervus elaphus*) in Europe. Quat Sci Rev 27:714–733
Sommer RS, Fahlke J, Schmölcke U, Benecke N, Zachos FE (2009) Quaternary history of the European roe deer (*Capreolus capreolus*). Mammal Rev 38:1–16
Sommer RS, Benecke N, Lougas L, Nelle O, Schmölcke U (2011) Holocene colonization pattern of the wild horse (*Equus ferus*) in Europe: a matter of landscape openness? J Quat Sci 26:805–812
Sommer RS, Kalbe J, Ekström J, Benecke N, Liljegren R (2014) Range dynamics of the reindeer in Europe during the last 25,000 years. J Biogeogr 41:298–306
Sommer RS, Hegge C, Schmölcke U (2018) Lack of support for adaptation of post-glacial horses to woodlands in the Central European Lowlands. Nat Ecol Evol 2:582–583
Soubrier J, Gower G, Chen K, Richards SM, Llamas B, Mitchell KJ, Ho SYW, Kosintsev P, Lee MSY, Baryshnikov G, Bollongino R, Bover P, Burger J, Chivall D, Crégut-Bonnoure E, Decker JE, Doronichev VB, Douka K, Fordham DA, Fontana F, Fritz C, Glimmerveen J, Golovanova LV, Groves C, Guerreschi A, Haak W, Higham T, Hofman-Kamińska E, Immel A, Julien MA, Krause J, Krotova O, Langbein F, Larson G, Rohrlach A, Scheu A, Schnabel RD, Taylor JF, Tokarska M, Tosello G, van der Plicht J, van Loenen A, Vigne J-D, Wooley O, Orlando L, Kowalczyk R, Shapiro B, Cooper A (2016) Early cave art and ancient DNA record the origin of European bison. Nat Commun 7:1–7
Stewart JR, Cooper A (2008) Ice Age refugia and Quaternary extinctions: an issue of Quaternary evolutionary palaeoecology. Quat Sci Rev 27:2443–2448
Street M, Baales M (1999) Pleistocene/Holocene changes in the Rhineland fauna in a northwest European context. Archäologie in Eurasien 6:9–38
Stuart AJ (2015) Late Quaternary megafaunal extinctions on the continents: a short review. Geol J 50:338–363
Stuart AJ, Lister AM (2011) Extinction chronology of the cave lion *Panthera spelaea*. Quat Sci Rev 30:2329–2340
Stuart AJ, Lister AM (2012) Extinction chronology of the woolly rhinoceros *Coelodonta antiquitatis* in the context of Late Quaternary megafaunal extinctions in Northern Eurasia. Quat Sci Rev 51:1–17
Stuart AJ, Lister AM (2014) New radiocarbon evidence on the extirpation of the spotted hyaena (Crocuta crocuta (Erxl.)) in northern Eurasia. Quat Sci Rev 96:108–116
Stuart AJ, Kosintsev PA, Higham TFG, Lister AM (2004) Pleistocene to Holocene extinction dynamics in giant deer and woolly mammoth. Nature 431:684–689
Theuerkauf M, Joosten H (2012) Younger Dryas cold stage vegetation patterns of Central Europe – climate, soil and relief controls. Boreas 41:391–407
von Koenigswald W (1974) Die pleistozäne Fauna der Weinberghöhlen bei Mauern. In: von Koenigswald W, Müller-Beck H, Pressmar E (eds) Archäologie und Paläontologie in den Weinberghöhlen von Mauern (Bayern) Grabungen. Selbstverlag für Urgeschichte, Tübingen, pp 1937–1967
von Koenigswald W (1977) Die Säugetierfauna aus der Burghöhle Dietfurt. Kölner Jahrbuch für Vor- und Frühgeschichte 15:123–142
von Koenigswald W (1984a) Die jungpleistozäne Säugetierfauna der Spitzbubenhöhle. In: Hahn J (ed) Die steinzeitliche Besiedlung des Eselsburger Tales bei Heidenheim. Landesdenkmalamt Baden-Württemberg. Konrad Theiss Verlag, Stuttgart
von Koenigswald W (1984b) Die Säugetiere der steinzeitlichen Station Malerfels I. In: Hahn J (ed) Die steinzeitliche Besiedlung des Eselsburger Tales bei Heidenheim. Landesdenkmalamt Baden-Württemberg. Konrad Theiss Verlag, Stuttgart
von Koenigswald W (2002) Lebendige Eiszeit. Klima und Tierwelt im Wandel. Theiss Verlag, Darmstadt
Weninger B, Jöris O, Danzglocke U (2008) CalPal 2007. Cologne radiocarbon calibration and palaeoclimate research package. http://www.calpal.de
Wohlfarth B, Veres D, Ampel L, Lacourse T, Blaauw M, Preusser F, Andrieu-Ponel V, Keravis D, Lallier-Verges E, Björck S, Davies S, de Beaulieu JL, Risberg J, Hormes A, Kasper HU, Possnert G, Reille M, Thouveny N, Zander A (2008) Rapid ecosystem response to abrupt climate changes during the last glacial period in Western Europe, 40–16 kyr BP. Geology 36:407–410
Yalden DW (1999) The history of British mammals. T. and A. D. Poyser, London
Yannouli E, Trantalidou K (1999) The fallow deer (*Dama dama* Linnaeus, 1758): archaeological presence and representation in Greece. Archäologie in Eurasien 6:247–281

Management of European Mammals

6

Klaus Hackländer and Arie Trouwborst

Contents

Abstract

In this chapter, we aim to provide an introduction to strategies, policy, and law regarding the management of mammals in Europe. Covering this extensive and complex topic in one concise chapter is no easy task, and we pretend no more than to scratch the surface. We begin by offering a bird's-eye view of some salient features concerning mammal policy and law in Europe, namely, (1) the sheer diversity of approaches; (2) the overall dominance of the paradigm of human-wildlife coexistence; (3) the influence and sophistication of international legal frameworks for wildlife conservation; and (4) the issues raised by the remarkable recent comeback of many large animal species, including large carnivores. Subsequently, we introduce two of the international frameworks in some detail, i.e., the pan-European Convention on the Conservation of European Wildlife and Natural Habitats of 1979 (Bern Convention) and the wildlife legislation of the European Union (EU). We then discuss an interesting but challenging approach that has been emerging with regard to wildlife populations that are shared amongst several European countries, such as large carnivores. This approach consists of adjusting management to the scale of each wildlife population – including where this population is transboundary – rather than adjusting it to the scale of countries or other jurisdictional units.

K. Hackländer (✉)
Department of Integrative Biology and Biodiversity Research, Institute of Wildlife Biology and Game Management, University of Natural Resources and Life Sciences, Vienna (BOKU), Vienna, Austria
e-mail: klaus.hacklaender@boku.ac.at

A. Trouwborst
Tilburg Law School, Department of Public Law and Governance, Tilburg University, Tilburg, The Netherlands
e-mail: a.trouwborst@tilburguniversity.edu

K. Hackländer, F. E. Zachos (eds.), *Mammals of Europe - Past, Present, and Future*, Handbook of the Mammals of Europe, https://doi.org/10.1007/978-3-030-00281-7_4

Keywords

Mammals · Europe · Management · Conservation · Control · Habitats Directive · Bern Convention · Wildlife · Human dimensions · Legal aspects

Introduction

In Europe, the conservation and use of mammals has a long and rich history. This history has been influenced by an array of cultural, socioeconomic, and political factors that have varied over time and continue to vary between regions and between countries (Temple and Terry 2007; Apollonio et al. 2010; Putman et al. 2011; Putman and Apollonio 2014; Linnell et al. 2015; Trouwborst and Hackländer 2018). Today, Europe is composed of a considerable number of very diverse, sovereign countries, each of which has its own distinct wildlife laws and policies, embedded within and influenced by its own ecological, cultural, and socioeconomic context. For instance, concepts of wildlife ownership, the legality of commercial uses of wildlife, and societal attitudes on hunting, to name a few variables, will often vary from one country to the next. A representative illustration is offered by the supplementary feeding of red deer (*Cervus elaphus*) which takes place in many parts of Europe, especially during winter, in order to increase population density, prevent mortality, and keep deer from over-wintering in areas where their presence is undesirable. In certain countries (parts of Austria, Croatia, Czech Republic, parts of Germany, Romania, and Slovakia), the provision of winter fodder to deer is actually a legal obligation of hunters; in certain other countries, it is a voluntary practice; whereas in the Netherlands and parts of Switzerland, supplementary feeding of deer is illegal (Linnell et al. 2015). Thus, when exploring wildlife law and policy in Europe, diversity as such stands out as a characteristic feature. Amidst this diversity, however, several common characteristics can be distinguished, some of which are introduced below.

In Europe, nature and culture are thoroughly intertwined, and coexistence rather than separation of humans and wildlife is an influential paradigm (Chapron et al. 2014; Boitani and Linnell 2015; Boitani and Sutherland 2015; Linnell et al. 2015; Carter and Linnell 2016). Almost everywhere, the landscape is dominated by human land-use, often manifestly so (Boitani and Sutherland 2015). In fact, many protected areas in Europe consist of intensively managed (private) land, and many recent conservation actions have targeted biodiversity on arable farmland (Boitani and Sutherland 2015). The practice of mammal management in Europe has thus been marked by rather blurred boundaries between nature and culture, between public and private land, between protected areas and the wider landscape, and indeed between wild and domestic animals (Linnell et al. 2015). Despite the continued predominance of the coexistence paradigm, more dualistic ideals, such as wilderness and rewilding, have been on the rise recently (Linnell et al. 2015). At any rate, conservation policies and laws in Europe tend to reflect a balance struck between ecological, socioeconomic, and cultural values. This will be illustrated below when discussing the Bern Convention and EU nature conservation law.

Without a doubt, one of the most striking features of wildlife policy and law in Europe as compared to other continents is the existence of comparatively sophisticated international legal frameworks for wildlife conservation, which exercise a notable influence in practice (Bowman et al. 2010; Fleurke and Trouwborst 2014). Foremost amongst these are the aforementioned Bern Convention and the EU Directive 92/43 on the Conservation of Natural Habitats and of Wild Fauna and Flora of 1992 (Habitats Directive). The Bern Convention and the Habitats Directive set out legally binding obligations for contracting countries concerning the protection of wildlife and the conservation of wildlife habitats. These obligations reflect a delicate balance struck between, on the one hand, a desire to cater for the ecological, cultural, and socioeconomic differences between different countries and localities and, on the other hand, a desire to provide common, overarching safeguards for the conservation of European biodiversity. This biodiversity is considered a European heritage of which individual countries are the custodians. The obligations thus provide certain minimum conservation standards to be observed across Europe or the EU. As long as countries abide by these minimum requirements, they are otherwise free to shape their own wildlife laws and policies (Trouwborst et al. 2017).

Other relevant international frameworks include the treaty regimes addressing wildlife conservation in two major European mountain ranges, the Alps and the Carpathians (see the Protocol to the Alpine Convention relating to the Conservation of Nature and the Countryside of 1994, and the Protocol to the Framework Convention on the Protection and Sustainable Development of the Carpathians on Conservation and Sustainable Use of Biological and Landscape Diversity of 2008). Invasive alien species and trade in wildlife are subject to specific EU legislation (Regulation 1143/2014 on the Prevention and Management of the Introduction and Spread of Invasive Alien Species and, inter alia, Regulation 338/97 on the Protection of Species of Wild Fauna and Flora by Regulating Trade therein). Finally, international cooperation for the conservation and management of various mammal species, particularly bats, cetaceans, and seals, has taken shape within the framework of the Bonn Convention on the Conservation of Migratory Species and a number of species-specific treaties adopted under the Bonn Convention's umbrella – namely, the Agreement on the Conservation of Seals in the Wadden Sea of 1990; the Agreement on the Conservation of Populations of European Bats of 1991 (EUROBATS); the Agreement on the Conservation of Small Cetaceans in the Baltic, North East Atlantic, Irish, and North Seas of 1992 (ASCOBANS); and the Agreement on the Conservation of Cetaceans of the Black Seas, Mediterranean and Contiguous Atlantic Area of 1996 (ACCOBAMS).

As may be expected in a densely populated continent with a long history of dynamic human development, biodiversity in Europe is under pressure. Many mammal species are in decline and have an unfavorable conservation status (European Environment Agency 2015). One in six of Europe's mammal species is under threat (Temple and Terry 2007). Threats include habitat loss and fragmentation, overexploitation, pollution, invasive alien species, and climate change. Farmland species such as the European hare (*Lepus europaeus*) or the Common hamster (*Cricetus cricetus*) are particularly hard hit, as they struggle to cope with rapidly changing agricultural practices and intensification across the continent (Hackländer and Schai-Braun 2017; Nechay 2000). Yet, the news is not all bad. In particular, recent decades have witnessed a spectacular comeback of a substantial number of large mammal and bird species in Europe (Deinet et al. 2013). This large wildlife comeback is driven by a combination of factors, including favorable habitat changes, shifting societal values, and improved legal protection. Mammals portraying population increases and range extensions include grey seal (*Halichoerus grypus*), harbor seal (*Phoca vitulina*), Eurasian beaver (*Castor fiber*), and an impressive range of ungulates: wild boar (*Sus scrofa*), roe deer (*Capreolus capreolus*), red deer, Eurasian moose (*Alces alces*), ibex (*Capra ibex* and *Capra pyrenaica*), chamois (*Rupicapra rupicapra* and *Rupicapra pyrenaica*), and the European bison (*Bison bonasus*).

Especially remarkable, however, is the recovery of Europe's guild of larger terrestrial carnivore species, namely, brown bear (*Ursus arctos*), wolf (*Canis lupus*), golden jackal (*Canis aureus*), Eurasian lynx (*Lynx lynx*), Iberian lynx (*Lynx pardinus*), and wolverine (*Gulo gulo*). The three biggest predators – bear, wolf, and Eurasian lynx – historically occupied most of the European continent, but then progressively disappeared from large parts of their ranges, mainly due to persistent prosecution by man. In recent decades, due to a mix of beneficial land use changes, recovering prey populations, changing societal attitudes, and legal protection, large carnivore populations have been recovering and reoccupying former ranges, with numbers rebounding to approximately 17,000 bears, 12,000 wolves, and 9,000 lynx at present (Chapron et al. 2014). All mainland European countries except for Belgium,

Luxembourg, the Netherlands, and small countries like Liechtenstein, Monaco, etc. currently have a permanent and reproducing occurrence of at least one large carnivore species (Chapron et al. 2014). Whereas this is an evident success story from a biodiversity conservation point of view, the coexistence of apex predators with humans remains a challenging affair, due to livestock depredation, competition with hunters, public safety concerns, and other kinds of conflicts with human interests. Especially fierce controversies have tended to ignite when carnivores return to (parts of) countries from which they had been absent for a long time.

That large carnivores have become one of the hottest topics in (inter)national European wildlife policy and law may be illustrated furthermore with reference to the plethora of current legal issues concerning these species. For instance, it is not always immediately apparent what the legal status is of large carnivores expanding their populations into countries from which they had disappeared long ago (Trouwborst 2010, 2014b, 2018), or where they did not previously occur. The latter situation has materialized with regard to the golden jackal, as this species is undergoing a remarkable northward and westward range extension beyond its traditional area of occurrence in southeastern Europe (Arnold et al. 2012). Jackals have already been observed as far north as the Baltic countries and as far northwest as Denmark and the Netherlands. This has led to confusion as to the legal status of the species in such countries where it is a newcomer, in particular, the issue whether to welcome the species as a dynamic part of Europe's native fauna or to treat it as an alien species to be curbed (Trouwborst et al. 2015). Further issues include the inadequate enforcement, in some regions, of European prohibitions on killing large carnivores (López-Bao et al. 2015); the legal status of wolf-dog hybrids (Trouwborst 2014a); the use of management zoning as a large carnivore management tool (Trouwborst 2014b, 2018); the uncertain geographic boundaries between different legal protection regimes in parts of Europe, particularly Spain (Trouwborst 2014c); the legality of border fences bisecting carnivore habitat (Trouwborst et al. 2016); and the question whether large carnivore populations should be managed at the national level or at the – often transboundary – population level (Trouwborst 2014b; Epstein et al. 2016; Trouwborst et al. 2017), a question revisited below.

The Bern Convention: A Pan-European Framework for Mammal Management

The Bern Convention is administered by the Council of Europe, an intergovernmental organization with a broader geographic scope than the EU. The Convention has 51 contracting parties, which include all 28 EU member states and the EU itself, as well as a range of non-EU countries across Eurasia, and even four African states (Burkina Faso, Morocco, Senegal, Tunisia). Within EU countries the Bern Convention is applied to mammals primarily through the Habitats Directive (see Epstein 2014). The objectives of the Convention are "to conserve wild flora and fauna and their natural habitats, especially those species and habitats whose conservation requires the co-operation of several States, and to promote such co-operation," whereby special emphasis is given to "endangered and vulnerable species" (Article 1). The Convention sets out various obligations to achieve these objectives. We concisely present some of the most significant ones, while referring readers interested in more elaborate analyses to other literature, such as Bowman et al. (2010), Lasén Díaz (2010), and Fleurke and Trouwborst (2014).

An important general obligation which applies to all wildlife requires parties to "take requisite measures to maintain the population of wild flora and fauna at, or adapt it to, a level which corresponds in particular to ecological, scientific and cultural requirements, while taking account of economic and recreational requirements and the sub-species, varieties or forms at risk locally" (Article 2). What this "level" amounts to exactly is not defined in more detail, but it appears safe to assume that species should at a minimum be conserved with a view to keeping them away from a threatened status on the IUCN Red Lists

Table 1 Bern Convention Appendices and Habitats Directive Annexes of relevance to mammals

Bern Convention	Appendix II	Strictly protected fauna species
	Appendix III	Protected fauna species
Habitats Directive	Annex I	Habitat types requiring designation of special areas of conservation
	Annex II	Species requiring designation of special areas of conservation
	Annex III	Criteria for selecting special areas of conservation
	Annex IV	Species in need of strict protection
	Annex V	Species which may be subject to management measures

(Bowman et al. 2010). Article 3 then calls for national conservation policies, "with particular attention to endangered and vulnerable species, especially endemic ones, and endangered habitats," as well as committing parties to "undertake" to "have regard to the conservation of wild flora and fauna" in their "planning and development policies."

These general provisions are accompanied by specific obligations concerning the conservation of habitats and the generic protection of species. These are directed principally, but not exclusively, at plant and animal species listed in three appendices to the Convention (see Table 1).

Appendix II lists several hundred "strictly protected fauna species," including a range of mammal taxa (e.g., Pyrenean desman (*Galemys pyrenaicus*), all Microchiroptera except common pipistrelle (*Pipistrellus pipistrellus*), woolly dormouse (*Dryomys laniger*), tiger (*Panthera tigris*), Abbruzzo chamois (*R. rupicapra ornata*), white-beaked dolphin (*Lagenorhynchus albirostris*)). Appendix III lists "protected fauna species" (without the adverb "strictly") and also covers a significant number of mammal taxa, including some commonly occurring species like the European hedgehog (*Erinaceus europaeus*), European hare (*Lepus europaeus*), or Eurasian badger (*Meles meles*).

As regards habitat protection, the Convention stipulates in Article 4 that for *all* wild flora and fauna species, each party "shall take appropriate and necessary legislative and administrative measures to ensure the conservation" of their habitats, with particular consideration for the habitats of species specified in Appendices I and II, and for "endangered natural habitats." The parties "in their planning and development policies shall have regard to the conservation requirements" of the areas thus protected, "so as to avoid or minimize as far as possible any deterioration of such areas." These general requirements of Article 4 have been elaborated through a series of decisions by the Standing Committee, the Bern Convention's main treaty body in which all parties are represented and which meets annually. Of particular significance are the decisions concerning the Network of Areas of Special Conservation Interest, or "Emerald Network" (e.g., Standing Committee Rec. No. 16 (1989)). In particular, contracting parties are requested to designate areas of special conservation interest (ASCI) pursuant to specific criteria as part of this Emerald Network and to ensure that the necessary conservation measures are taken for each ASCI.

As regards generic species protection, each party to the Convention "shall take appropriate and necessary legislative and administrative measures to ensure the special protection of the wild fauna species specified in Appendix II," and similar measures to "ensure the protection" (without the adjective "special") of animal species from Appendix III (Articles 6–7). With respect to Appendix II fauna, Article 6, in unequivocal terms, requires parties to prohibit:

a) all forms of deliberate capture and keeping and deliberate killing;
b) the deliberate damage to or destruction of breeding or resting sites;
c) the deliberate disturbance of wild fauna, particularly during the period of breeding, rearing and hibernation insofar as disturbance would be significant in relation to the objectives of this Convention;
d) . . .;
e) the possession of and internal trade in these animals, alive or dead, including stuffed animals and any readily recognizable part of derivative thereof, where this would contribute to the effectiveness of the provisions of this article.

Furthermore, any exploitation of animals belonging to Appendix III species "shall be regulated in order to keep the populations out of danger," for instance by closing certain seasons for hunting or through regulation of trade (Article 7). Regarding animals from Appendices II and III, parties "shall prohibit the use of all indiscriminate means of capture and killing and the use of all means capable of causing local disappearance of, or serious disturbance to, populations," and especially the means specified in Appendix IV to the Convention (Article 8). Snares, poisoned baits, and (semi) automatic weapons figure among the prohibited items included in this fourth Appendix. Article 9 of the Convention allows parties to grant exemptions from the above prohibitions when three cumulative conditions are met: (i) there is "no other satisfactory solution"; (ii) "the exception will not be detrimental to the survival of the population concerned"; and (iii) the exception is made for one of the following purposes:

- for the protection of flora and fauna;
- to prevent serious damage to crops, livestock, forests, fisheries, water and other forms of property;
- in the interests of public health and safety, air safety or other overriding public interests;
- for the purposes of research and education, of repopulation, of reintroduction and for the necessary breeding;
- to permit, under strictly supervised conditions, on a selective basis and to a limited extent, the taking, keeping or other judicious exploitation of certain wild animals and plants in small numbers.

Regarding the reintroduction of native species of flora and fauna, Article 11 of the Convention specifies that when certain conditions are met, parties are to "encourage" such reintroduction "when this would contribute to the conservation of an endangered species." At the same time, parties are to "strictly control the introduction of non-native species."

On the whole, the strictness and unqualified nature of the obligations laid down in the Bern Convention stand out. As Bowman et al. (2010, 198) put it, "almost every one of its provisions is mandatory, as opposed to being couched in the hortatory language favored by many conservation treaties." The Convention does, however, expressly allow states when they become a party – and only then – to submit reservations regarding specific species from Appendices II–III or means/methods from Appendix IV (Article 22). These species and/or means are then exempted from the obligations of the party concerned under the Convention. Over 20 states parties have availed themselves of this possibility. Many reservations concern large carnivores. For instance, reservations regarding the wolf – a strictly protected Appendix II species – have been submitted by Bulgaria, the Czech Republic, Finland, Latvia, Lithuania, Macedonia, Poland, Slovakia, Slovenia, Spain, Turkey, and Ukraine.

The obligations of parties under the Convention have come to be informed by a significant body of guidance laid down in Recommendations adopted by the Standing Committee. Such guidance is not itself legally binding but does carry considerable interpretive value as regards the application of the Convention's binding provisions. A relevant example concerns the European Charter on Biodiversity and Hunting of 2007 (Brainerd 2007). The Standing Committee has recommended the parties to the Convention to incorporate the 12 principles set out and elaborated in the Charter within their hunting policies "so as to ensure that hunting is carried out in a sustainable way" (Standing Committee Rec. No. 128 (2007)). Said principles, to provide some examples, call on states to "ensure that harvest is ecologically sustainable"; "maintain wild populations of indigenous species with adaptive gene pools"; "maintain environments that support healthy and robust populations of harvestable species"; "encourage acceptance by society of sustainable, consumptive use as a conservation tool"; and – quite sensibly – "ensure that regulations are understandable and respected" (Principles 3, 4, 5, 12, and 2, respectively).

Compliance by contracting parties with their Convention obligations is promoted inter alia through compulsory reporting by the parties and through the so-called case file procedure, whereby alleged breaches of the Convention are brought to the attention of the Standing Committee, often by NGOs. The Committee may then examine the potential violation, including through an on-the-spot appraisal, and as appropriate recommend a particular course of action to the contracting party involved to resolve the situation.

Whereas the influence of the Bern Convention on wildlife conservation and management in Europe should not be exaggerated, there are plenty of indications to suggest that the Convention is exercising a meaningful influence on domestic policies and decisions affecting wildlife (Fleurke and Trouwborst 2014).

The EU Habitats Directive

The Habitats Directive has a more restricted geographic scope than the Bern Convention, as it binds only the current 28 EU member states. Again, our analysis of the Directive must be concise for reasons of space. More elaborate commentaries have been provided by Bowman et al. (2010), García Ureta (2010), Jones (2012), Fleurke and Trouwborst (2014), and Born et al. (2015).

In terms of substance, there is a notable albeit incomplete overlap between the Bern Convention and the Habitats Directive. The latter targets the conservation of 220 habitat types and approximately 1,000 species of (non-avian) fauna and flora listed in its annexes. It does not cover common mammal species like, for instance, roe deer, wild boar, red fox (*Vulpes vulpes*), or European hare. The Directive aims for the maintenance or achievement of a "favourable conservation status" for the habitats and species it covers, in order to contribute to biodiversity conservation in Europe (Article 2). It stipulates in general terms that all measures taken by EU member states pursuant to the Directive "shall be designed to maintain or restore, at favorable conservation status, natural habitats and species of wild fauna and flora of Community interest" (Article 2). Unlike the rather undefined conservation level to be achieved under the Bern Convention, the Habitats Directive defines the concept of "favourable conservation status" in some detail. For instance, the status of a species is deemed favorable when, inter alia, the species "is maintaining itself on a long-term basis as a viable component of its natural habitats" and "there is, and will probably continue to be, a sufficiently large habitat to maintain its populations on a long-term basis" (Article 1). Like the Bern Convention, the Directive contains various specific obligations concerning the protection of areas and the generic protection of species.

The area protection provisions in the Habitats Directive, however, are more sophisticated than the Bern Convention's Article 4. Under the Directive, specific obligations regarding the designation and protection of sites are to be taken with regard to natural habitat types listed in the Directive's Annex I and species listed in Annex II (Article 4). After a multiple-stage procedure involving both the member states and the European Commission, sites of importance for these habitats and species are to be designated as "special areas of conservation" (SACs), which then become part of the EU network of protected sites denominated "Natura 2000." Natura 2000 constitutes the contribution of the EU and its member states to the Bern Convention's Emerald Network, mentioned above. Importantly, member states are to employ ecological criteria only for the selection and delimitation of sites under the Birds and Habitats Directives. The selection of SACs is based on a set of criteria provided in Annex III of the Habitats Directive, which include the representativeness of habitats, their degree of conservation, and the relative population density of species. Significantly, the Natura 2000 network presently covers nearly one-fifth (18.17%) of the total EU land territory (Sundseth 2018).

As regards the conservation of Natura 2000 sites, Article 6 of the Habitats Directive requires member states to take "the necessary conservation measures" which "correspond to the ecological requirements" of the habitats and species protected within SACs. Furthermore, they "shall take appropriate steps to avoid," in SACs, "the deterioration of natural habitats." Concrete plans and projects which might harm the protected nature within designated SACs are subject to a restrictive authorization scheme, laid down in Articles 6(3)–(4) of the Habitats Directive –

literally reproduced here because of its considerable impact in practice:

> (3) Any plan or project not directly connected with or necessary to the management of the site but likely to have a significant effect thereon, either individually or in combination with other plans or projects, shall be subject to appropriate assessment of its implications for the site in view of the site's conservation objectives. In light of the conclusions of the assessment of the implications for the site and subject to the provisions of paragraph 4, the competent authorities shall agree to the plan or project only after having ascertained that it will not adversely affect the integrity of the site concerned [. . .].
>
> (4) If, in spite of a negative assessment of the implications for the site and in the absence of alternative solutions, a plan or project must nevertheless be carried out for imperative reasons of overriding public interest, including those of a social or economic nature, the Member State shall take all compensatory measures necessary to ensure that the overall coherence of Natura 2000 is protected.

The Court of Justice of the EU (CJEU), which is the ultimate authority regarding the interpretation of EU legislation, has developed extensive case law regarding the Nature Directives' (i.e., Habitats Directive and Birds Directive) rules on site designation and protection. Throughout this case law, the Court has tended to interpret the rules involved in such a way as to maximize their effectiveness in light of the Directives' conservation objectives. For example, the Court has clarified that considerations of an economic nature, or concerning expected future management difficulties, may play no part in the site designation process (e.g., Case C-355/90, Commission v Spain, [1993] ECR I-4221; Case C-3/96, Commission v Netherlands, [1998] ECR I-3031). Another example is an (in)famous judgment concerning the assessment and authorization of plans and projects, issued by the CJEU in a case involving fisheries in the Dutch Wadden Sea. Interpreting Article 6(3), the Court determined that plans or projects may in principle be authorized only "where no reasonable scientific doubt remains as to the absence" of harmful impacts (Case C-127/02, Waddenvereniging, [2004] ECR I-7405, par. 61).

Regarding generic species protection, Annex IV of the Habitats Directive lists species that are to be strictly protected, whereas a more flexible regime applies to another set of species included in Annex V (see Table 1). For Annex IV mammal taxa (e.g., Sicilian shrew (*Crocidura sicula*), Egyptian fruit bat (*Rousettus aegyptiacus*), Caucasian squirrel (*Sciurus anomalus*), Arctic fox (*Alopex lagopus*), wild goat (*Capra aegagrus*), harbor porpoise (*Phocoena phocoena*)), the Directive requires member states to "take the requisite measures to establish a system of strict protection" (Articles 12 and 13). Regarding such fauna, Article 12 requires the establishment of prohibitions on inter alia the killing, capturing, and disturbing of individual animals, and on the "deterioration or destruction of breeding sites or resting places." CJEU case law makes it clear that member states must not only prohibit the acts in question but also take all measures necessary to ensure that the prohibitions in question are not violated in practice (e.g., Case C-103/00, Commission v Greece, [2002] ECR I-1147). Exemptions from these prohibitions may only be granted when (all of) three conditions set out in Article 16 are met, conditions which roughly approximate those set out in the Bern Convention, mentioned previously. A specific duty to monitor and address "incidental capture and killing" of Annex IV taxa – e.g., in road traffic – is also laid down in Article 12. Member states are to take the conservation measures necessary to ensure that such killing does not have a "significant negative impact" on the species involved.

Annex V covers "Animal and plant species of community interest whose taking in the wild and exploitation may be subject to management measures." Taxa such as Eurasian pine marten (*Martes martes*), mountain hare (*Lepus timidus*), or ibex (*Capra ibex*) are listed in Annex V. Their status leaves competent authorities with significantly more leeway regarding conservation and management actions. Article 14 of the Habitats Directive enumerates measures that may be applied by member states to regulate the exploitation of Annex V populations, e.g., closed seasons and license systems. Nonetheless, these are presented as options rather than obligations. The discretionary room for member states regarding Annex V species is not unlimited, however. They are bound by a general obligation to ensure a favorable conservation status for such species (García Ureta 2010). Two further obligations apply to Annex IV and Annex V species alike. First, Article 11 obliges member states to guarantee that surveillance of the conservation status of such species "is undertaken systematically and on a permanent basis" (Case C-6/04, Commission v United Kingdom, [2005] ECR I-9017, par. 68). Second, Article 15 of the Directive outlaws the use of certain means and modes of capture and killing, including poison(ed baits), (semi)automatic weapons, and all other "indiscriminate means capable of causing local disappearance of, or serious disturbance to, populations." Exceptions to allow such means may only be made when the three aforementioned conditions of Article 16 are met.

All in all, the Habitats Directive imposes clear limits on member states' discretion regarding mammal conservation and management. Moreover, the nature of EU law allows individual stakeholders to directly invoke the Directives before national courts. Compliance is also enhanced through the enforcement machinery at the EU level, where the European Commission operates as watchdog and has the option of submitting alleged cases of non-compliance by member states with the Nature Directives to the CJEU. Persistent non-compliance can result in the imposition by the Court of major financial penalties. As a result, the Nature Directives are highly effective when compared to other international legal instruments for biodiversity conservation (Verschuuren 2003; Bowman et al. 2010; Fleurke and Trouwborst 2014; Born et al. 2015). Many species have profited from the protection of their habitat and/or the restrictions placed on their exploitation (Deinet et al. 2013). The wolf is a good example where the Habitats Directive's impact on the conservation status of species can be clearly appreciated (Fleurke and Trouwborst 2014; Chapron et al. 2014).

However, the implementation of the Nature Directives has hitherto not been flawless either, due inter alia to the Directives' delayed or faulty transposition into national law, insufficient law enforcement at the domestic level – for instance, regarding the illegal killing of protected species, such as wolves in southern Spain (Trouwborst 2014c; López-Bao et al. 2015) – flawed and inconsistent application of procedures at the EU level (Krämer 2009), and confusion concerning the interpretation of certain terms and concepts contained in the Directives' provisions (Jones 2012; Wandesforde-Smith and Watts 2014; Born et al. 2015; Epstein et al. 2016; Trouwborst et al. 2017).

The EU LIFE Programme: A Flagship Approach to Mammal Conservation

L'Instrument Financier pour l'Environnement (LIFE) is the European Union's financial instrument to support environmental and nature conservation as well as climate action projects. Projects are not restricted to EU countries and might include EU candidates and EU neighboring countries. Since 1992, LIFE has co-financed (on average 50% of the total costs) more than 4500 projects. About every seventh LIFE project (more than 180) focused on or benefited mammals. Of these, 105 specifically targeted mammalian taxa. About 50% of the projects aimed at large carnivore conservation, followed by Chiroptera (Temple and Terry 2007; Salsi 2011). Apart from that, projects for Mediterranean monk seals (*Monachus monachus*) or Siberian flying squirrels (*Pteromys volans*) have been conducted. Examples of actions taken include the development of species action plans, habitat restoration, habitat conservation, and reintroductions. Interestingly, almost half of the mammal projects were located in Spain and Italy, partly due to the fact that LIFE projects are restricted to Natura 2000 networks.

Examples of the numerous success stories of LIFE projects include the recovery of Iberian lynx in Portugal and Spain, mainly by recovering European wild rabbit (*Oryctolagus cuniculus*) populations, the main prey of Iberian lynx. Nearly a dozen LIFE projects targeted cetaceans, such as bottlenose dolphins (*Tursiops truncatus*), and focused on monitoring habitat use and conflicts with fisheries. In Poland, LIFE ensured the survival of the heaviest terrestrial mammal in Europe, the European bison (*Bison bonasus*) by habitat improvements, habitat protection, and contracts with local farmers maintaining meadows that are favorable for bison dispersal. For European mink (*Mustela lutreola*), reintroductions took place in the course of LIFE projects in Estonia, while in Spain the invasive American mink (*Neovison vison*) was culled in European mink habitats.

However, more than 70 mammal taxa protected at EU level by the Bern Convention or Habitat Directive have not been targeted by the LIFE programme, despite their unfavorable conservation status. These include, for example, Canary shrew (*Crocidura canariensis*), Cabrera's vole (*Microtus cabrerae*), wolverine (*Gulo gulo*), or Balkan chamois (*R. rupicapra balcanica*) (Salsi 2011). In 2013, the EU decided to extend the program (now called LIFE+). For the 2014–2020 funding period, LIFE will contribute approximately €3.4 billion to the protection of the environment and climate, of which €2.6 billion are dedicated for the subprogram Environment which includes thematic priorities for nature and biodiversity as well as water including the marine environment. Current projects to be found in the LIFE database deal again with European bison, brown bears, and European mink, but also include a reintroduction project of West Indian manatee (*Trichechus manatus*) in the French overseas region Guadeloupe.

Management of Transboundary Wildlife Populations: A Special Challenge

From a wildlife conservation and management perspective, it is preferable to adjust pertinent policies and laws to the spatial scale of a wildlife population, rather than adjusting it to biologically meaningless political and administrative boundaries. This presents special challenges, however, in particular where a population straddles the territories of various countries. The actual implementation of conservation and management at the transboundary population level is a complex and challenging affair. Certain recent developments in Europe are at the forefront in this regard and merit discussion in the present context. We provide such discussion below, albeit concisely, in respect of large carnivores. For more elaborate analyses, we refer to other literature on the topic (Linnell et al. 2008; Linnell and Boitani 2012; Blanco 2013; Epstein 2013; Trouwborst 2014b, 2015; Selier et al. 2016; Trouwborst et al. 2017).

Wolves, bears, lynx, and wolverines naturally occur at comparably low population densities and move over vast areas, with individuals typically having yearly home ranges varying from 100 to 1,000 km^2. This makes the population concept for large carnivores meaningless on small scales of less than many thousands, or even many tens of thousands, of square kilometers. In Europe, therefore, their populations tend to stretch across many jurisdictional boundaries, including international frontiers (Linnell et al. 2001, 2008; Linnell and Boitani 2012; Chapron et al. 2014). The four largest predator species presently occur in 33 distinct (sub)populations across Europe, 28 of which are shared between two or more countries: 8 out of 10 wolf populations are transboundary, 8 out of 10 bear populations, 10 out of 11 Eurasian lynx populations, and 2 out of 2 wolverine populations (Kaczensky et al. 2013; Chapron et al. 2014). To effectively manage these wide-ranging species at a population level, the need for transboundary coordination is thus especially strong. Some basic elements of the envisioned cross-border approach are described in the following statement by Linnell and Boitani (2012, 84) regarding wolves:

> The first step that is required is to move away from viewing wolf distribution within the arbitrary lines on maps that national or provincial borders represent and to look at the actual distribution. The resulting view is one of a 'meta-population like' structure where demographic viability is achievable in many regional units that have a more or less continuous distribution of wolves (populations). It is crucial that these populations are managed as biological units – with the administrative bodies (be they intra- or inter-national) that share a population coordinating their activities to ensure that

their independent actions enhance rather than hinder each other.

Both the Bern Convention and the Habitats Directive set out obligations concerning the generic protection of the four large carnivore species, and the protection of their habitat. However, these obligations target the countries involved individually. No provision is made for concerted conservation actions tailored to transboundary wildlife populations, notwithstanding a generally phrased obligation in the Bern Convention for parties to "cooperate whenever appropriate and in particular where this would enhance the effectiveness of measures taken under other articles of this Convention" (Article 11). Besides, the legal regimes applicable to the different species under these instruments vary from country to country, due to reservations submitted under the Bern Convention and country-specific differences established under the Habitats Directive (Trouwborst 2010). To give an example, under the Bern Convention, depending on the party concerned, the wolf is a "strictly protected fauna species" under Appendix II, a "protected fauna species" under Appendix III, or neither. Similar differences apply under the Habitats Directive and to some of the other species involved. The situation is compounded further by the fact that not all Bern Convention parties are also EU member states. The fragmented legal regime for wolves in Europe is mapped in Fig. 1.

This fragmentation of the European legal landscape in respect of large carnivores adds to the urgency of transboundary cooperation at the population level.

In the face of this situation, both the Standing Committee of the Bern Convention and the European Commission expressly advocate a transboundary population-level approach to large carnivore conservation and management. Of particular interest is a detailed guidance document developed by the Large Carnivore Initiative for Europe (a specialist group of the IUCN Species Survival Commission) under contract from the European Commission (Linnell et al. 2008). These "Guidelines for Population Level Management Plans for Large Carnivores in Europe" (Carnivore Guidelines) call for the adoption of a population-level management plan, by the competent authorities of all countries involved, for each large carnivore population, and set out detailed instructions in this regard (Linnell et al. 2008). When endorsing the Carnivore Guidelines in 2008, the European Commission stated that "it is difficult, if not impossible, for one Member State to manage and protect its large carnivores in the absence of concerted and convergent actions being taken by its neighbours" (European Commission 2008). Specifically, the Commission held that "effective management of large carnivore populations which are shared between Member States can only be achieved through shared and coordinated management plans as described in the[se] guidelines," considering the Carnivore Guidelines to represent "best practice" when it comes to the application of the Habitats Directive to large carnivores (European Commission 2008). The Standing Committee of the Bern Convention has likewise called on parties to the Convention to "reinforce cooperation with neighbouring states in view of adopting harmonized policies towards management of shared populations of large carnivores, taking into account the best practice in the field of management of populations of large carnivores," under express reference to the Carnivore Guidelines (Rec. No. 137 (2008), par. 1; see also Rec. No. 115 (2005)).

The Carnivore Guidelines provide a detailed template setting out the ingredients that each transboundary management plan should contain (Linnell et al. 2008). According to this template, the objectives for the population concerned should be "specific and measurable," encompassing concrete goals in terms of numbers, range, and other parameters such as harvest rates, damage levels, and poaching levels, "that can be used to measure the success of management actions" (Linnell et al. 2008, par. 2.2). These goals ought to be "distributed in space" between the various administrative units involved, such as "countries, states, counties, wildlife management units or protected areas" (par. 2.2). Regarding specific actions, the template stresses that it is crucial for the removal of animals to be "coordinated between all management units that share a

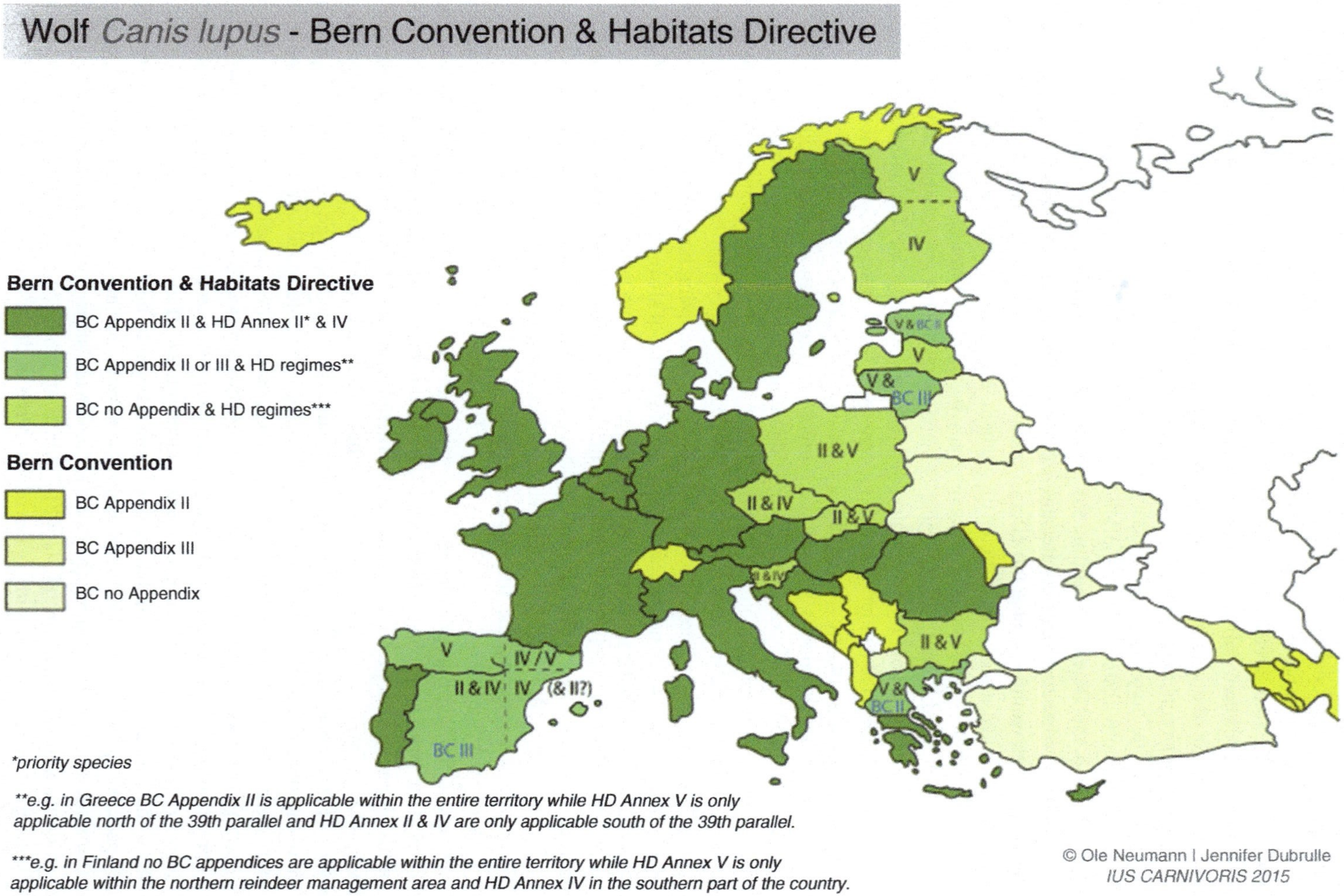

Fig. 1 Legal status of wolves under Bern Convention (BC) and Habitats Directive (HD). (Reproduced with permission from Ole Neumann and Jennifer Dubrulle (IUS CARNIVORIS 2015))

population," based on a predetermined "population level limit for the number of individuals that can be removed per year" (par. 3.6). Serious attention should, furthermore, be paid to ensuring connectivity within the population as well as with neighboring populations (par. 2.2.5 and 3.2). A final point we single out here is that each plan should indicate any "changes in legislation that are needed to bring about the population level management plan" (par. 3.3). Although the Carnivore Guidelines generally refer to population-level management "plans," they clarify that the transboundary cooperation concerned may take any of various shapes, as long as it adequately serves its purpose. It could involve a legally binding agreement, but this is not a strict requirement. The arrangement involved needs to be sufficiently flexible to adjust to future developments regarding the population concerned but also sufficiently formal and high-profile to warrant its actual observation by the governmental actors involved (Linnell et al. 2008; Trouwborst 2010).

Currently, several attempts have been taken to speed up this process. For example, in June 2014, the EU Commission initiated the EU platform on Coexistence between People and Large Carnivores, a grouping of organizations representing different interest groups, including the European Landowners' Organization ELO, Reindeer herders, the European Federation of Associations for Hunting & Conservation FACE, the International Council for Game and Wildlife Conservation CIC, the IUCN, WWF, and the largest European organization of protected areas, EUROPARC Federation. In regular workshops the member organizations discuss "ways and means to minimize, and wherever possible find solutions to, conflicts between human interests and the presence of large carnivore species, by exchanging knowledge and by working together in an open-ended, constructive, and mutually respectful way." Moreover, in October 2016 the Framework Convention on the Protection and Sustainable Development of the Carpathians (Carpathian Convention) signed a Memorandum of Cooperation with CIC, in which they agreed to "enhance the capacity of countries in the Carpathian region in conserving and sustainably managing wildlife resources, thereby contributing to biodiversity conservation, food safety and nutrition (food security), wildlife, livestock, human health, and social and economic development." Carpathian-wide management strategies for large carnivores will be the main focus for this starting cooperation. Despite these attempts, the pace at which this population-level approach is actually being implemented by European countries in respect of large carnivores is slow. Notwithstanding a number of promising initiatives, the first fully-fledged transboundary population-level management plan is still to be formalized (Blanco 2013). It would, however, be unrealistic to expect huge strides in this respect. As Linnell and Boitani (2012) reflect, "it is only a few decades since wolves changed their official status from vermin to conservation icons [and therefore] not surprising that the process takes time and is stormy."

References

Apollonio M, Andersen R, Putman R (eds) (2010) European ungulates and their management in the 21st century. Cambridge University Press, Cambridge

Arnold J, Humer A, Heltai M, Murariu D, Spassov N, Hackländer K (2012) Current status and distribution of golden jackals (*Canis aureus* L., 1758) in Europe. Mammal Rev 42:1–11

Blanco JC (2013) Towards a population level approach for the management of large carnivores in Europe: challenges and opportunities. European Commission, Brussels

Boitani L, Linnell JDC (2015) Bringing large mammals back: large carnivores in Europe. In: Pereira HM, Navarro LM (eds) Rewilding European landscapes. Springer, Cham, pp 67–84

Boitani L, Sutherland WJ (2015) Introduction special section: conservation in Europe as a model for emerging conservation issues globally. Conserv Biol 29:975–977

Born C, Cliquet A, Schoukens H, Misonne D, van Hoorick G (eds) (2015) The Habitats Directive in its EU environmental law context: European nature's best hope? Routledge, London

Bowman M, Redgwell C, Davies P (2010) Lyster's international wildlife law, 2nd edn. Cambridge University Press, Cambridge

Brainerd S (2007) European Charter on hunting and biodiversity. Council of Europe, Strasbourg

Carter NH, Linnell JDC (2016) Co-adaptation is key to coexisting with large carnivores. Trends in Ecology & Evolution 31:575–578

Chapron G, Kaczensky P, Linnell JDC, von Arx M, Huber D, Andrén H, López-Bao JV, Adamec M, Álvares F, Anders O, Balčiauskas L, Balys V, Bedő P, Bego F, Blanco JC, Breitenmoser U, Brøseth H, Bufka L, Bunikyte R, Ciucci P, Dutsov A, Engleder T, Fuxjäger C, Groff C, Holmala K, Hoxha B, Iliopoulos Y, Ionescu O, Jeremić J, Jerina K, Kluth G, Knauer F, Kojola I, Kos I, Krofel M, Kubala J, Kunovac S, Kusak J, Kutal M, Liberg O, Majić A, Männil P, Manz R, Marboutin E, Marucco F, Melovski D, Mersini K, Mertzanis Y, Mysłajek RW, Nowak S, Odden J, Ozolins J, Palomero G, Paunović M, Persson J, Potočnik H, Quenette P-Y, Rauer G, Reinhardt I, Rigg R, Ryser A, Salvatori V, Skrbinšek T, Stojanov A, Swenson JE, Szemethy L, Trajçe A, Tsingarska-Sedefcheva E, Váňa M, Veeroja R, Wabakken P, Wölfl M, Wölfl S, Zimmermann F, Zlatanova D, Boitani L (2014) Recovery of large carnivores in Europe's modern human-dominated landscapes. Science 346:1517–1519

Deinet S, Ieronymidou C, McRae L, Burfield IJ, Foppen RP, Collen B, Böhm M (2013) Wildlife comeback in Europe: the recovery of selected mammal and bird species. Zoological Society of London, Birdlife International and European Bird Census Council, London

Epstein Y (2013) Population based species management across legal boundaries: the Bern Convention, Habitats Directive, and the gray wolf in Scandinavia. Georget Int Environ Law Rev 25:549–597

Epstein Y (2014) The Habitats Directive and Bern Convention: synergy and dysfunction in public international and EU law. Georget Int Environ Law Rev 26:139–173

Epstein Y, López-Bao JV, Chapron G (2016) A legal-ecological understanding of favorable conservation status for species in Europe. Conserv Lett 9:81–88

European Commission (2008) Note to the guidelines for population level management plans for large carnivores. European Commission, Brussels

European Environment Agency (2015) State of nature in the EU: results from reporting under the Nature Directives 2007–2012. Publications Office of the European Union, Luxembourg

Fleurke FM, Trouwborst A (2014) European regional approaches to the transboundary conservation of biodiversity: the Bern Convention and the EU Birds and Habitats Directives. In: Kotze L, Marauhn T (eds) Transboundary governance of biodiversity. Martinus Nijhoff, Leiden/Boston, pp 128–162

García Ureta A (2010) Derecho europeo de la biodiversidad: aves silvestres, hábitats y especies de flora y fauna. Iustel, Madrid

Hackländer K, Schai-Braun SC (2017) *Lepus europaeus* Pallas, 1778 European hare. In: Smith AT, Johnston CH, Alves PC, Hackländer K (eds) Lagomorphs: pikas, rabbits and hares of the world. John Hopkins University Press, Baltimore, pp 187–190

Jones G (ed) (2012) The Habitats Directive: a developer's obstacle course? Hart, Oxford/Portland

Kaczensky P, Chapron G, von Arx M, Huber D, Andrén H, Linnell JDC (eds) (2013) Status, management and distribution of large carnivores – bear, lynx, wolf and wolverine – in Europe. European Commission, Brussels

Krämer L (2009) The European Commission's opinions under Article 6(4) of the Habitats Directive. J Environ Law 21:59–85

Lasén Díaz C (2010) The Bern Convention: 30 years of nature conservation in Europe. Rev Eur Community Int Environ Law 19:185–196

Linnell JDC, Boitani L (2012) Building biological realism into wolf management policy: the development of the population approach in Europe. Hystrix 23:80–91

Linnell JDC, Andersen R, Kvam T, Andrén H, Liberg O, Odden J, Moa P (2001) Home range size and choice of management strategy for lynx in Scandinavia. Environ Manag 27:869–879

Linnell JDC, Salvatori V, Boitani L (2008) Guidelines for population level management plans for large carnivores in Europe. European Commission, Brussels

Linnell JDC, Kaczensky P, Wotschikowsky U, Lescureux N, Boitani L (2015) Framing the relationship between people and nature in the context of European conservation. Conserv Biol 29:978–985

López-Bao JV, Blanco JC, Rodríguez A, Godinho R, Sazatornil V, Alvares F, García EJ, Llaneza L, Rico M, Cortés Y, Palacios V, Chapron G (2015) Toothless wildlife protection laws. Biodivers Conserv 24:2105–2108

Nechay G (2000) Status of hamsters: *Cricetus cricetus*, *Cricetus migratorius*, *Mesocricetus newtoni* and other hamster species in Europe. Nature and environment series, no. 106. Council of Europe Publishing, Strasbourg

Putman R, Apollonio M (2014) Behaviour and management of European ungulates. Whittles Publishing, Dunbeath

Putman R, Apollonio M, Andersen R (2011) Ungulate management in Europe: problems and practices. Cambridge University Press, Cambridge

Salsi A (ed) (2011) LIFE and European mammals. Improving their conservation status. Office for Official Publications of the European Communities, Luxembourg

Selier SAJ, Slotow R, Blackmore A, Trouwborst A (2016) The legal challenges of transboundary wildlife management at the population level: the case of a trilateral elephant population in southern Africa. J Int Wildl Law Policy 19:101

Sundseth K (2018) 25 years of LIFE – supporting nature in Europe! Natura 2000 Newsletter 43. Office for Official Publications of the European Communities, Luxembourg

Temple HJ, Terry A (2007) The status and distribution of European mammals. Office for Official Publications of the European Communities, Luxembourg

Trouwborst A (2010) Managing the carnivore comeback: international and EU species protection law and the return of lynx, wolf and bear to Western Europe. J Environ Law 22:347–372

Trouwborst A (2014a) Exploring the legal status of wolf-dog hybrids and other dubious animals: international and EU law and the wildlife conservation problem of hybridization with domestic and alien species. Rev Eur Comp Int Environ Law 23:111–124

Trouwborst A (2014b) Living with success – and with wolves: addressing the legal issues raised by the unexpected homecoming of a controversial carnivore. Eur Energy Environ Law Rev 23:89–101

Trouwborst A (2014c) The EU Habitats Directive and wolf conservation and management on the Iberian Peninsula: a legal perspective. Galemys 26:15–30

Trouwborst A (2015) Global large carnivore conservation and international law. Biodivers Conserv 24: 1567–1588

Trouwborst A (2018) Wolves not welcome? Zoning for large carnivore conservation and management under the Bern Convention and EU Habitats Directive. Rev Eur Comp Int Environ Law 27:306–319

Trouwborst A, Hackländer K (2018) Wildlife policy and law in Europe. In: Leopold BC, Kessler WB, Cummins JL (eds) North American wildlife policy and law. Boone and Crocket Club, Missoula, pp 425–445

Trouwborst A, Krofel M, Linnell JDC (2015) Legal implications of range expansions in a terrestrial carnivore: the case of the golden jackal (*Canis aureus*) in Europe. Biodivers Conserv 24:2593–2610

Trouwborst A, Fleurke F, Dubrulle J (2016) Border fences and their impacts on large carnivores, large herbivores and biodiversity: an international wildlife law perspective. Rev Eur Comp Int Environ Law 25:291–306

Trouwborst A, Boitani L, Linnell JDC (2017) Interpreting 'favourable conservation status' for large carnivores in Europe: how many are needed and how many are wanted? Biodivers Conserv 26:37–61

Verschuuren JM (2003) Effectiveness of nature protection legislation in the EU and the US: the Birds and Habitats Directives and the Endangered Species Act. Yearb Eur Environ Law 3:305–328

Wandesforde-Smith G, Watts NSJ (2014) Wildlife conservation and protected areas: politics, procedure, and the performance of failure under the EU Birds and Habitats Directives. J Int Wildl Law Policy 17:62–80

Index

K. Hackländer, F. E. Zachos (eds.), *Mammals of Europe - Past, Present, and Future*,
Handbook of the Mammals of Europe, https://doi.org/10.1007/978-3-030-00281-7

The manufacturer's authorised representative in the EU is Springer Nature Customer Service Centre GmbH, Europaplatz 3, 69115 Heidelberg, Germany. If you have any concerns regarding our products, please contact ProductSafety@springernature.com

Printed and bound by CPI Group (UK) Ltd, Croydon, CR0 4YY

30/09/2026

02225469-0007